THE HOLOGRAPHIC UNIVERSE

Also by Michael Talbot

NONFICTION
Mysticism and the New Physics
Beyond the Quantum
Your Past Lives: A Reincarnation Handbook

FICTION
The Delicate Dependency
The Bog
Night Things

THE HOLOGRAPHIC UNIVERSE

THE
HOLOGRAPHIC
UNIVERSE

MICHAEL TALBOT

HarperCollins*Publishers*

THE HOLOGRAPHIC UNIVERSE. Copyright © 1991 by Michael Talbot. All
rights reserved. Printed in the United States of America. No part of this
book may be used or reproduced in any manner whatsoever without written
permission except in the case of brief quotations embodied in critical articles and reviews. For information address HarperCollins Publishers, 10
East 53rd Street, New York, NY 10022.

FIRST EDITION

Designed by Helene Berinsky

LIBRARY OF CONGRESS CATALOG CARD NUMBER 90-55555

91 92 93 94 95 CC/RRD 10 9 8 7 6 5 4 3 2

For Alexandra, Chad, Ryan,
Larry Joe, and Shawn,
with love

The new data are of such far-reaching relevance that they could revolutionize our understanding of the human psyche, of psychopathology, and of the therapeutic process. Some of the observations transcend in their significance the framework of psychology and psychiatry and represent a serious challenge to the current Newtonian-Cartesian paradigm of Western science. They could change drastically our image of human nature, of culture and history, and of reality.

> —Dr. Stanislav Grof on
> holographic phenomena in
> *The Adventure of Self-Discovery*

Contents

Acknowledgments

Writing is always a collaborative effort and many people have contributed to the production of this book in various ways. It is not possible to name them all, but a few who deserve special mention include:

David Bohm, Ph.D., and Karl Pribram, Ph.D., who were generous with both their time and their ideas, and without whose work this book would not have been written.

Barbara Brennan, M.S., Larry Dossey, M.D., Brenda Dunne, Ph.D., Elizabeth W. Fenske, Ph.D., Gordon Globus, Jim Gordon, Stanislav Grof, M.D., Francine Howland, M.D., Valerie Hunt, Ph.D., Robert Jahn, Ph.D., Ronald Wong Jue, Ph.D., Mary Orser, F. David Peat, Ph.D., Elizabeth Rauscher, Ph.D., Beatrice Rich, Peter M. Rojcewicz, Ph.D., Abner Shimony, Ph.D., Bernie S. Siegel, M.D., T.M. Srinivasan, M.D., Whitley Strieber, Russell Targ, William A. Tiller, Ph.D., Montague Ullman, M.D., Lyall Watson, Ph.D., Joel L. Whitton, M.D., Ph.D., Fred Alan Wolf, Ph.D., and Richard Zarro, who were also all generous with their time and ideas.

Carol Ann Dryer, for her friendship, insight, and support, and for unending generosity when it comes to sharing her profound talent.

Kenneth Ring, Ph.D., for hours of fascinating conversation and for introducing me to the writings of Henry Corbin.

Stanley Krippner, Ph.D., for taking the time to call me or drop me a note whenever he came across any new leads on the holographic idea.

Terry Oleson, Ph.D., for his time and for kindly allowing me to use his diagram of the "little man in the ear."

Michael Grosso, Ph.D., for thought-provoking conversation and for helping me track down several obscure reference works on miracles.

Brendan O'Regan of the Institute of Noetic Sciences, for his important contributions to the subject of miracles and for helping me track down information on the same.

My longtime friend Peter Brunjes, Ph.D., for using his university connections to help me obtain several difficult-to-find reference works.

Judith Hooper, for loaning me numerous books and articles from her own extensive collection of materials on the holographic idea.

Susan Cowles, M.S., of the Museum of Holography in New York for helping me search out illustrations for the book.

Kerry Brace, for sharing his thoughts on the holographic idea as it applies to Hindu thinking, and from whose writings I have borrowed the idea of using the hologram of Princess Leia from the movie *Star Wars* to open the book.

Marilyn Ferguson, the founder of the *Brain/Mind Bulletin,* who was one of the first writers to recognize and write about the importance of the holographic theory, and who also was generous with her time and thought. The observant reader will notice that my summary of the view of the universe that arises when one considers Bohm and Pribram's conclusions in tandem, at the end of Chapter Two, is actually just a slight rephrasing of the words Ferguson uses to summarize the same sentiment in her bestselling book *The Aquarian Conspiracy.* My inability to come up with a different and better way to summarize the holographic idea should be viewed as a testament to Ferguson's clarity and succinctness as a writer.

The staff at the American Society for Psychical Research for assistance in tracking down references, resources, and the names of pertinent individuals.

Martha Visser and Sharon Schuyler for their help in researching the book.

Ross Wetzsteon of the *Village Voice,* who asked me to write the article that started it all.

Claire Zion of Simon & Schuster, who first suggested that I write a book on the holographic idea.

Lucy Kroll and Barbara Hogenson for being the best agents possible.

Lawrence P. Ashmead of HarperCollins for believing in the book, and John Michel for his gentle and insightful editing.

If there is anyone that I have inadvertently left out, please forgive me. To all, both named and unnamed, who have helped me give birth to this book, my heartfelt thanks.

THE HOLOGRAPHIC UNIVERSE

Introduction

In the movie *Star Wars,* Luke Skywalker's adventure begins when a beam of light shoots out of the robot Artoo Detoo and projects a miniature three-dimensional image of Princess Leia. Luke watches spellbound as the ghostly sculpture of light begs for someone named Obi-wan Kenobi to come to her assistance. The image is a *hologram,* a three-dimensional picture made with the aid of a laser, and the technological magic required to make such images is remarkable. But what is even more astounding is that some scientists are beginning to believe the universe itself is a kind of giant hologram, a splendidly detailed illusion no more or less real than the image of Princess Leia that starts Luke on his quest.

Put another way, there is evidence to suggest that our world and everything in it—from snowflakes to maple trees to falling stars and spinning electrons—are also only ghostly images, projections from a level of reality so beyond our own it is literally beyond both space and time.

The main architects of this astonishing idea are two of the world's most eminent thinkers: University of London physicist David Bohm, a protégé of Einstein's and one of the world's most respected quantum physicists; and Karl Pribram, a neurophysiologist at Stanford University and author of the classic neuropsychological textbook *Languages of the Brain.* Intriguingly, Bohm and Pribram arrived at their conclusions independently and while working from two very different directions. Bohm became convinced of the universe's holographic nature only after years of dissatisfaction with standard theories' inability to explain all of the phenomena encountered in quantum physics. Pribram became convinced because of the failure of standard theories of the brain to explain various neurophysiological puzzles.

However, after arriving at their views, Bohm and Pribram quickly realized the holographic model explained a number of other mysteries as well, including the apparent inability of any theory, no matter how comprehensive, ever to account for all the phenomena encountered in nature; the ability of individuals with hearing in only one ear to determine the direction from which a sound originates; and our ability to recognize the face of someone we have not seen for many years even if that person has changed considerably in the interim.

But the most staggering thing about the holographic model was that it suddenly made sense of a wide range of phenomena so elusive they generally have been categorized outside the province of scientific understanding. These include telepathy, precognition, mystical feelings of oneness with the universe, and even psychokinesis, or the ability of the mind to move physical objects without anyone touching them.

Indeed, it quickly became apparent to the ever growing number of scientists who came to embrace the holographic model that it helped explain virtually all paranormal and mystical experiences, and in the last half-dozen years or so it has continued to galvanize researchers and shed light on an increasing number of previously inexplicable phenomena. For example:

- In 1980 University of Connecticut psychologist Dr. Kenneth Ring proposed that near-death experiences could be explained by the holographic model. Ring, who is president of the International Association for Near-Death Studies, believes such experiences, as well as death itself, are really nothing more than the shifting of a person's consciousness from one level of the hologram of reality to another.

- In 1985 Dr. Stanislav Grof, chief of psychiatric research at the Maryland Psychiatric Research Center and an assistant professor of psychiatry at the Johns Hopkins University School of Medicine, published a book in which he concluded that existing neurophysiological models of the brain are inadequate and only a holographic model can explain such things as archetypal experiences, encounters with the collective unconscious, and other unusual phenomena experienced during altered states of consciousness.

- At the 1987 annual meeting of the Association for the Study of Dreams held in Washington, D.C., physicist Fred Alan Wolf delivered a talk in which he asserted that the holographic model explains lucid dreams (unusually vivid dreams in which the dreamer realizes he or she is awake). Wolf believes such dreams are actually visits to parallel realities, and the holographic model will ultimately allow us to develop a "physics of consciousness" which will enable us to begin to explore more fully these other-dimensional levels of existence.

- In his 1987 book entitled *Synchronicity: The Bridge Between Matter and Mind,* Dr. F. David Peat, a physicist at Queen's University in Canada, asserted that synchronicities (coincidences that are so unusual and so psychologically meaningful they don't seem to be the result of chance alone) can be explained by the holographic model. Peat believes such coincidences are actually "flaws in the fabric of reality." They reveal that our thought processes are much more intimately connected to the physical world than has been hitherto suspected.

These are only a few of the thought-provoking ideas that will be explored in this book. Many of these ideas are extremely controversial. Indeed, the holographic model itself is highly controversial and is by no means accepted by a majority of scientists. Nonetheless, and as we shall see, many important and impressive thinkers do support it and believe it may be the most accurate picture of reality we have to date.

The holographic model has also received some dramatic experimental support. In the field of neurophysiology numerous studies have corroborated Pribram's various predictions about the holographic nature of memory and perception. Similarly, in 1982 a landmark experiment performed by a research team led by physicist Alain Aspect at the Institute of Theoretical and Applied Optics, in Paris, demonstrated that the web of subatomic particles that compose our physical universe—the very fabric of reality itself—possesses what appears to be an undeniable "holographic" property. These findings will also be discussed in the book.

In addition to the experimental evidence, several other things add weight to the holographic hypothesis. Perhaps the most important considerations are the character and achievements of the two men who originated the idea. Early in their careers, and before the holographic model was even a glimmer in their thoughts, each amassed accomplishments that would inspire most researchers to spend the rest of their academic lives resting on their laurels. In the 1940s Pribram did pioneering work on the limbic system, a region of the brain involved in emotions and behavior. Bohm's work in plasma physics in the 1950s is also considered landmark.

But even more significantly, each has distinguished himself in another way. It is a way even the most accomplished men and women can seldom call their own, for it is measured not by mere intelligence or even talent. It is measured by courage, the tremendous resolve it takes to stand up for one's convictions even in the face of overwhelming opposition. While he was a graduate student, Bohm did doctoral work with Robert Oppenheimer. Later, in 1951, when Oppenheimer came under the perilous scrutiny of Senator Joseph McCarthy's Committee on Un-American Activities, Bohm was called to testify against him and refused. As a result he lost his job at Princeton and never again taught in the United States, moving first to Brazil and then to London.

Early in his career Pribram faced a similar test of mettle. In 1935 a Portuguese neurologist named Egas Moniz devised what he believed was the perfect treatment for mental illness. He discovered that by boring into an individual's skull with a surgical pick and severing the prefrontal cortex from the rest of the brain he could make the most troublesome patients docile. He called the procedure a *prefrontal lobotomy*, and by the 1940s it had become such a popular medical technique that Moniz was awarded the Nobel Prize. In the 1950s the procedure's popularity continued and it became a tool, like the McCarthy hearings, to stamp out cultural undesirables. So accepted was its use for this purpose that the surgeon Walter Freeman, the most outspoken advocate for the procedure in the United States, wrote unashamedly that lobotomies "made good American citizens" out of society's misfits, "schizophrenics, homosexuals, and radicals."

During this time Pribram came on the medical scene. However, unlike many of his peers, Pribram felt it was wrong to tamper so recklessly with the brain of another. So deep were his convictions that while working as a young neurosurgeon in Jacksonville, Florida, he opposed the accepted medical wisdom of the day and refused to allow any lobotomies to be performed in the ward he was overseeing. Later at Yale he maintained his controversial stance, and his then radical views very nearly lost him his job.

Bohm and Pribram's commitment to stand up for what they believe in, regardless of the consequences, is also evident in the holographic model. As we shall see, placing their not inconsiderable reputations behind such a controversial idea is not the easiest path either could have taken. Both their courage and the vision they have demonstrated in the past again add weight to the holographic idea.

One final piece of evidence in favor of the holographic model is the paranormal itself. This is no small point, for in the last several decades a remarkable body of evidence has accrued suggesting that our current understanding of reality, the solid and comforting sticks-and-stones picture of the world we all learned about in high-school science class, is wrong. Because these findings cannot be explained by any of our standard scientific models, science has in the main ignored them. However, the volume of evidence has reached the point where this is no longer a tenable situation.

To give just one example, in 1987, physicist Robert G. Jahn and clinical psychologist Brenda J. Dunne, both at Princeton University, announced that after a decade of rigorous experimentation by their Princeton Engineering Anomalies Research Laboratory, they had accumulated unequivocal evidence that the mind can psychically interact with physical reality. More specifically, Jahn and Dunne found that through mental concentration alone, human beings are able to affect the way certain kinds of machines operate. This is an astounding finding and one that cannot be accounted for in terms of our standard picture of reality.

It can be explained by the holographic view, however. Conversely, because paranormal events cannot be accounted for by our current scientific understandings, they cry out for a new way of looking at the universe, a new scientific paradigm. In addition to showing how the holographic model can account for the paranormal, the book will also examine how mounting evidence in favor of the paranormal in turn actually seems to necessitate the existence of such a model.

The fact that the paranormal cannot be explained by our current scientific worldview is only one of the reasons it remains so controversial. Another is that psychic functioning is often very difficult to pin down in the lab, and this has caused many scientists to conclude it therefore does not exist. This apparent elusiveness will also be discussed in the book.

An even more important reason is that contrary to what many of us have come to believe, science is not prejudice-free. I first learned this a number of years ago when I asked a well-known physicist what he thought about a particular parapsychological experiment. The physicist (who had a reputation for being skeptical of the paranormal) looked at me and with great authority said the results revealed "no evidence of any psychic functioning whatsoever." I had not yet seen the results, but because I respected the physicist's intelligence and reputation, I accepted his judgment without question. Later when I examined the results for myself, I was stunned to discover the experiment had produced very striking evidence of psychic ability. I realized then that even well-known scientists can possess biases and blind spots.

Unfortunately this is a situation that occurs often in the investigation of the paranormal. In a recent article in *American Psychologist*, Yale psychologist Irvin L. Child examined how a well-known series of ESP dream experiments conducted at the Maimonides Medical Center in Brooklyn, New York, had been treated by the scientific establishment. Despite the dramatic evidence supportive of ESP uncovered by the experimenters, Child found their work had been almost completely ignored by the scientific community. Even more distressing, in the handful of scientific publications that had bothered to comment on the experiments, he found the research had been so "severely distorted" its importance was completely obscured.[1]

How is this possible? One reason is science is not always as objective as we would like to believe. We view scientists with a bit of awe, and when they tell us something we are convinced it must be true. We forget they are only human and subject to the same religious, philosophical, and cultural prejudices as the rest of us. This is unfortunate, for as this book will show, there is a great deal of evidence that the universe encompasses considerably more than our current worldview allows.

But why is science so resistant to the paranormal in particular? This is a more difficult question. In commenting on the resistance he experienced to his own unorthodox views on health, Yale surgeon Dr. Bernie S. Siegel, author of the best-selling book *Love, Medicine, and Miracles*, asserts that it is because people are addicted to their beliefs. Siegel says this is why when you try to change someone's belief they act like an addict.

There seems to be a good deal of truth to Siegel's observation, which perhaps is why so many of civilization's greatest insights and advances have at first been greeted with such passionate denial. We *are* addicted to our beliefs and we *do* act like addicts when someone tries to wrest from us the powerful opium of our dogmas. And since Western science has devoted several centuries to not believing in the paranormal, it is not going to surrender its addiction lightly.

I am lucky. I have always known there was more to the world than is generally accepted. I grew up in a psychic family, and from an early age I experienced firsthand many of the phenomena that will be talked about in this book. Occasionally, and when it is relevant to the topic being discussed, I will relate a few of my own experiences. Although they can only be viewed as anecdotal evidence, for me they have provided the most compelling proof of all that we live in a universe we are only just beginning to fathom, and I include them because of the insight they offer.

Lastly, because the holographic concept is still very much an idea in the making and is a mosaic of many different points of view and pieces of evidence, some have argued that it should not be called a model or theory until these disparate points of view are integrated into a more unified whole. As a result, some researchers refer to the ideas as the *holographic paradigm.* Others prefer *holographic analogy, holographic metaphor,* and so on. In this book and for the sake of diversity I have employed all of these expressions, including *holographic model* and *holographic theory,* but do not mean to imply that the holographic idea has achieved the status of a model or theory in the strictest sense of these terms.

In this same vein it is important to note that although Bohm and Pribram are the originators of the holographic idea, they do not embrace all of the views and conclusions put forward in this book. Rather, this is a book that looks not only at Bohm and Pribram's theories, but at the ideas and conclusions of numerous researchers who have been influenced by the holographic model and who have interpreted it in their own sometimes controversial ways.

Throughout this book I also discuss various ideas from quantum physics, the branch of physics that studies subatomic particles (electrons, protons, and so on). Because I have written on this subject before, I am aware that some people are intimidated by the term *quantum physics* and are afraid they will not be able to understand its concepts. My experience has taught me that even those who do not know any mathematics are able to understand the kinds of ideas from physics that are touched upon in this book. You do not even need a background in science. All you need is an open mind if you happen to glance at a page and see a scientific term you do not know. I have kept such terms down to a minimum, and on those occasions when it was necessary to use one, I always explain it before continuing on with the text.

So don't be afraid. Once you have overcome your "fear of the water," I think you'll find swimming among quantum physics' strange and fascinating ideas much easier than you thought. I think you'll also find that pondering a few of these ideas might even change the way you look at the world. In fact, it is my hope that the ideas contained in the following chapters *will* change the way you look at the world. It is with this humble desire that I offer this book.

PART I

A REMARKABLE
NEW VIEW
OF REALITY

Sit down before fact like a little child, and be prepared to give up every preconceived notion, follow
humbly wherever and to whatever abyss Nature
leads, or you shall learn nothing.

—T. H. Huxley

1

The Brain as Hologram

It isn't that the world of appearances is wrong; it isn't that there *aren't* objects out there, at one level of reality. It's that if you penetrate through and look at the universe with a holographic system, you arrive at a different view, a different reality. And that other reality can explain things that have hitherto remained inexplicable scientifically: paranormal phenomena, synchronicities, the apparently meaningful coincidence of events.

—Karl Pribram
in an interview in *Psychology Today*

The puzzle that first started Pribram on the road to formulating his holographic model was the question of how and where memories are stored in the brain. In the early 1940s, when he first became interested in this mystery, it was generally believed that memories were localized in the brain. Each memory a person had, such as the memory of the last time you saw your grandmother, or the memory of the fragrance of a gardenia you sniffed when you were sixteen, was believed to have a specific location somewhere in the brain cells. Such memory traces were called *engrams,* and although no one knew what an engram was made of—whether it was a neuron or perhaps even a special kind of molecule—most scientists were confident it was only a matter of time before one would be found.

There were reasons for this confidence. Research conducted by Canadian neurosurgeon Wilder Penfield in the 1920s had offered convincing evidence that specific memories did have specific locations in the brain. One of the most unusual features of the brain is that the object itself doesn't sense pain directly. As long as the scalp and skull have been deadened with a local anesthetic, surgery can be performed on the brain of a fully conscious person without causing any pain.

In a series of landmark experiments, Penfield used this fact to his advantage. While operating on the brains of epileptics, he would electrically stimulate various areas of their brain cells. To his amazement he found that when he stimulated the temporal lobes (the region of the brain behind the temples) of one of his fully conscious patients, they reexperienced memories of past episodes from their lives in vivid detail. One man suddenly relived a conversation he had had with friends in South Africa; a boy heard his mother talking on the telephone and after several touches from Penfield's electrode was able to repeat her entire conversation; a woman found herself in her kitchen and could hear her son playing outside. Even when Penfield tried to mislead his patients by telling them he was stimulating a different area when he was not, he found that when he touched the same spot it always evoked the same memory.

In his book *The Mystery of the Mind,* published in 1975, just shortly before his death, he wrote, "It was evident at once that these were not dreams. They were electrical activations of the sequential record of consciousness, a record that had been laid down during the patient's earlier experience. The patient 're-lived' all that he had been aware of in that earlier period of time as in a moving-picture 'flashback.' "[1]

From his research Penfield concluded that everything we have ever experienced is recorded in our brain, from every stranger's face we have glanced at in a crowd to every spider web we gazed at as a child. He reasoned that this was why memories of so many insignificant events kept cropping up in his sampling. If our memory is a complete record of even the most mundane of our day-to-day experiences, it is reasonable to assume that dipping randomly into such a massive chronicle would produce a good deal of trifling information.

As a young neurosurgery resident, Pribram had no reason to doubt Penfield's engram theory. But then something happened that was to change his thinking forever. In 1946 he went to work with the great neuropsychologist Karl Lashley at the Yerkes Laboratory of Primate Biology, then in Orange Park, Florida. For over thirty years Lashley had been involved in his own ongoing search for the elusive mechanisms responsible for memory, and there Pribram was able to witness the fruits of Lashley's labors firsthand. What was startling was that not only had Lashley failed to produce any evidence of the engram, but his research actually seemed to pull the rug out from under all of Penfield's findings.

What Lashley had done was to train rats to perform a variety of tasks, such as run a maze. Then he surgically removed various portions of their brains and retested them. His aim was literally to cut out the area of the rats' brains containing the memory of their maze-running ability. To his surprise he found that no matter what portion of their brains he cut out, he could not eradicate their memories. Often the rats' motor skills were impaired and they stumbled clumsily through the mazes, but even with massive portions of their brains removed, their memories remained stubbornly intact.

For Pribram these were incredible findings. If memories possessed specific locations in the brain in the same way that books possess specific locations on library shelves, why didn't Lashley's surgical plunderings have any effect on them? For Pribram the only answer seemed to be that memories were not localized at specific brain sites, but were somehow spread out or *distributed* throughout the brain as a whole. The problem was that he knew of no mechanism or process that could account for such a state of affairs.

Lashley was even less certain and later wrote, "I sometimes feel, in reviewing the evidence on the localization of the memory trace, that the necessary conclusion is that learning just is not possible at all. Nevertheless, in spite of such evidence against it, learning does sometimes occur."[2] In 1948 Pribram was offered a position at Yale, and before leaving he helped write up thirty years of Lashley's monumental research.

The Breakthrough

At Yale, Pribram continued to ponder the idea that memories were distributed throughout the brain, and the more he thought about it the more convinced he became. After all, patients who had had portions of their brains removed for medical reasons never suffered the loss of specific memories. Removal of a large section of the brain might cause a patient's memory to become generally hazy, but no one ever came out of surgery with any selective memory loss. Similarly, individuals who had received head injuries in car collisions and other accidents never forgot half of their family, or half of a novel they had read. Even removal of sections of the temporal lobes, the area of the brain that had figured so prominently in Penfield's research, didn't create any gaps in a person's memories.

Pribram's thinking was further solidified by his and other researchers' inability to duplicate Penfield's findings when stimulating brains other than those of epileptics. Even Penfield himself was unable to duplicate his results in nonepileptic patients.

Despite the growing evidence that memories were distributed, Pribram was still at a loss as to how the brain might accomplish such a seemingly magical feat. Then in the mid-1960s an article he read in *Scientific American* describing the first construction of a hologram hit him like a thunderbolt. Not only was the concept of holography dazzling, but it provided a solution to the puzzle with which he had been wrestling.

To understand why Pribram was so excited, it is necessary to understand a little more about holograms. One of the things that makes holography possible is a phenomenon known as interference. Interference is the crisscrossing pattern that occurs when two or more waves, such as waves of water, ripple through each other. For example, if you drop a pebble into a pond, it will produce a series of concentric waves that expands outward. If you drop two pebbles into a pond, you will get two sets of waves that expand and pass through one another. The complex arrangement of crests and troughs that results from such collisions is known as an interference pattern.

Any wavelike phenomena can create an interference pattern, including light and radio waves. Because laser light is an extremely pure, coherent form of light, it is especially good at creating interference patterns. It provides, in essence, the perfect pebble and the perfect pond. As a result, it wasn't until the invention of the laser that holograms, as we know them today, became possible.

A hologram is produced when a single laser light is split into two separate beams. The first beam is bounced off the object to be photographed. Then the second beam is allowed to collide with the reflected light of the first. When this happens they create an interference pattern which is then recorded on a piece of film (see fig. 1).

To the naked eye the image on the film looks nothing at all like the object photographed. In fact, it even looks a little like the concentric rings that form when a handful of pebbles is tossed into a pond (see fig. 2). But as soon as another laser beam (or in some instances just a bright light source) is shined through the film, a three-dimensional image of the original object reappears. The three-dimensionality of such images is often eerily convincing. You can actually walk around a holographic projection and view it from different angles as you would a real object. However, if you reach out and try to touch it, your hand will waft right through it and you will discover there is really nothing there (see fig. 3).

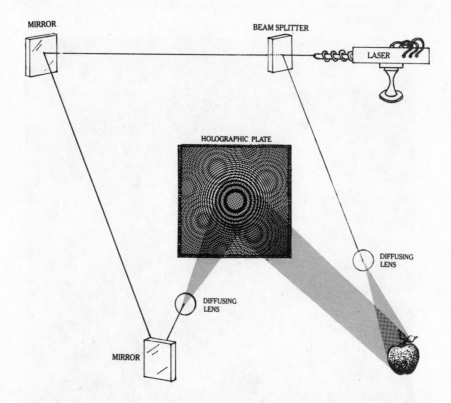

FIGURE 1. A hologram is produced when a single laser light is split into two separate beams. The first beam is bounced off the object to be photographed, in this case an apple. Then the second beam is allowed to collide with the reflected light of the first, and the resulting interference pattern is recorded on film.

Three-dimensionality is not the only remarkable aspect of holograms. If a piece of holographic film containing the image of an apple is cut in half and then illuminated by a laser, each half will still be found to contain the entire image of the apple! Even if the halves are divided again and then again, an entire apple can still be reconstructed from each small portion of the film (although the images will get hazier as the portions get smaller). Unlike normal photographs, every

FIGURE 2. A piece of holographic film containing an encoded image. To the naked eye the image on the film looks nothing like the object photographed and is composed of irregular ripples known as *interference patterns*. However, when the film is illuminated with another laser, a three-dimensional image of the original object reappears.

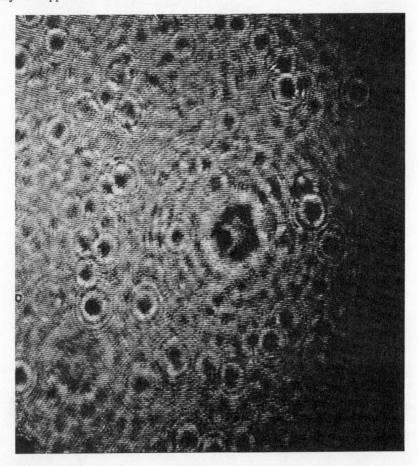

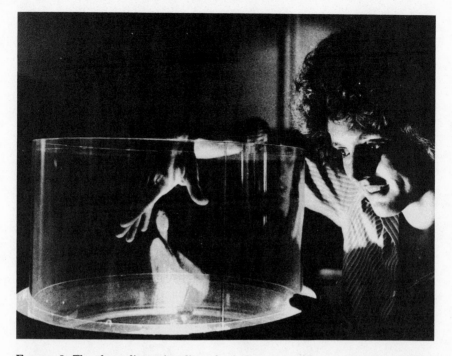

FIGURE 3. The three-dimensionality of a hologram is often so eerily convincing that you can actually walk around it and view it from different angles. But if you reach out and try to touch it, your hand will waft right through it. ["Celeste Undressed." Holographic stereogram by Peter Claudius, 1978. Photograph by Brad Cantos, collection of The Museum of Holography. Used by permission]

small fragment of a piece of holographic film contains all the information recorded in the whole (see fig. 4).*

This was precisely the feature that got Pribram so excited, for it offered at last a way of understanding how memories could be distributed rather than localized in the brain. If it was possible for every portion of a piece of holographic film to contain all the information necessary to create a whole image, then it seemed equally possible for every part of the brain to contain all of the information necessary to recall a whole memory.

*It should be noted that this astounding trait is common only to pieces of holographic film whose images are invisible to the naked eye. If you buy a piece of holographic film (or an object containing a piece of holographic film) in a store and can see a three-dimensional image in it without any special kind of illumination, do not cut it in half. You will only end up with pieces of the original image.

FIGURE 4. Unlike normal photographs, every portion of a piece of holographic film contains all of the information of the whole. Thus if a holographic plate is broken into fragments, each piece can still be used to reconstruct the entire image.

Vision Also Is Holographic

Memory is not the only thing the brain may process holographically. Another of Lashley's discoveries was that the visual centers of the brain were also surprisingly resistant to surgical excision. Even after removing as much as 90 percent of a rat's visual cortex (the part of the brain that receives and interprets what the eye sees), he found it could still perform tasks requiring complex visual skills. Similarly, research conducted by Pribram revealed that as much as 98 percent of a cat's optic nerves can be severed without seriously impairing its ability to perform complex visual tasks.[3]

Such a situation was tantamount to believing that a movie audience could still enjoy a motion picture even after 90 percent of the movie screen was missing, and his experiments presented once again a serious challenge to the standard understanding of how vision works. According to the leading theory of the day, there was a one-to-one correspondence between the image the eye sees and the way that image is represented in the brain. In other words, when we look at a square, it was believed the electrical activity in our visual cortex also possesses the form of a square (see fig. 5).

Although findings such as Lashley's seemed to deal a deathblow to this idea, Pribram was not satisfied. While he was at Yale he devised a series of experiments to resolve the matter and spent the next seven years carefully measuring the electrical activity in the brains of monkeys while they performed various visual tasks. He discovered that not only did no such one-to-one correspondence exist, but there wasn't even a discernible pattern to the sequence in which the electrodes fired. He wrote of his findings, "These experimental results are incompatible with a view that a photographic-like image becomes projected onto the cortical surface."[4]

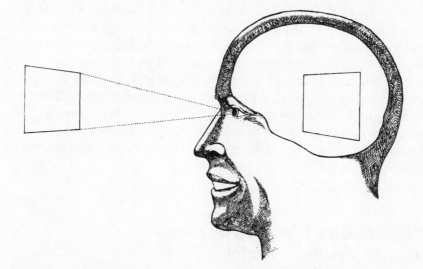

FIGURE 5. Vision theorists once believed there was a one-to-one correspondence between an image the eye sees and how that image is represented in the brain. Pribram discovered this is not true.

Once again the resistance the visual cortex displayed toward surgical excision suggested that, like memory, vision was also distributed, and after Pribram became aware of holography he began to wonder if it, too, was holographic. The "whole in every part" nature of a hologram certainly seemed to explain how so much of the visual cortex could be removed without affecting the ability to perform visual tasks. If the brain was processing images by employing some kind of internal hologram, even a very small piece of the hologram could still reconstruct the whole of what the eyes were seeing. It also explained the lack of any one-to-one correspondence between the external world and the brain's electrical activity. Again, if the brain was using holographic principles to process visual information, there would be no more one-to-one correspondence between electrical activity and images seen than there was between the meaningless swirl of interference patterns on a piece of holographic film and the image the film encoded.

The only question that remained was what wavelike phenomenon the brain might be using to create such internal holograms. As soon as Pribram considered the question he thought of a possible answer. It was known that the electrical communications that take place between the brain's nerve cells, or neurons, do not occur alone. Neurons possess branches like little trees, and when an electrical message reaches the end of one of these branches it radiates outward as does the ripple in a pond. Because neurons are packed together so densely, these expanding ripples of electricity—also a wavelike phenomenon—are constantly crisscrossing one another. When Pribram remembered this he realized that they were most assuredly creating an almost endless and kaleidoscopic array of interference patterns, and these in turn might be what give the brain its holographic properties. "The hologram was there all the time in the wave-front nature of brain-cell connectivity," observed Pribram. "We simply hadn't had the wit to realize it."[5]

Other Puzzles Explained by the Holographic Brain Model

Pribram published his first article on the possible holographic nature of the brain in 1966, and continued to expand and refine his ideas during the next several years. As he did, and as other researchers became aware of his theory, it was quickly realized that the distributed nature of memory and vision is not the only neurophysiological puzzle the holographic model can explain.

THE VASTNESS OF OUR MEMORY

Holography also explains how our brains can store so many memories in so little space. The brilliant Hungarian-born physicist and mathematician John von Neumann once calculated that over the course of the average human lifetime, the brain stores something on the order of 2.8×10^{20} (280,000,000,000,000,000,000) bits of information. This is a staggering amount of information, and brain researchers have long struggled to come up with a mechanism that explains such a vast capability.

Interestingly, holograms also possess a fantastic capacity for information storage. By changing the angle at which the two lasers strike a piece of photographic film, it is possible to record many different images on the same surface. Any image thus recorded can be retrieved simply by illuminating the film with a laser beam possessing the same angle as the original two beams. By employing this method researchers have calculated that a one-inch-square of film can store the same amount of information contained in fifty Bibles![6]

OUR ABILITY TO BOTH RECALL AND FORGET

Pieces of holographic film containing multiple images, such as those described above, also provide a way of understanding our ability to both recall and forget. When such a piece of film is held in a laser beam and tilted back and forth, the various images it contains appear and disappear in a glittering stream. It has been suggested that our ability to remember is analogous to shining a laser beam on such a piece of film and calling up a particular image. Similarly, when we are unable to recall something, this may be equivalent to shining various beams on a piece of multiple-image film, but failing to find the right angle to call up the image/memory for which we are searching.

ASSOCIATIVE MEMORY

In Proust's *Swann's Way* a sip of tea and a bite of a small scallop-shaped cake known as a *petite madeleine* cause the narrator to find himself suddenly flooded with memories from his past. At first he is puzzled, but then, slowly, after much effort on his part, he remembers that his aunt used to give him tea and madeleines when he was a little boy, and it is this association that has stirred his memory. We have all had similar experiences—a whiff of a particular food being prepared, or a glimpse of some long-forgotten object—that suddenly evoke some scene out of our past.

The holographic idea offers a further analogy for the associative tendencies of memory. This is illustrated by yet another kind of holographic recording technique. First, the light of a single laser beam is bounced off two objects simultaneously, say an easy chair and a smoking pipe. The light bounced off each object is then allowed to collide, and the resulting interference pattern is captured on film. Then, whenever the easy chair is illuminated with laser light and the light that reflects off the easy chair is passed through the film, a three-dimensional image of the pipe will appear. Conversely, whenever the same is done with the pipe, a hologram of the easy chair appears. So, if our brains function holographically, a similar process may be responsible for the way certain objects evoke specific memories from our past.

OUR ABILITY TO RECOGNIZE FAMILIAR THINGS

At first glance our ability to recognize familiar things may not seem so unusual, but brain researchers have long realized it is quite a complex ability. For example, the absolute certainty we feel when we spot a familiar face in a crowd of several hundred people is not just a subjective emotion, but appears to be caused by an extremely fast and reliable form of information processing in our brain.

In a 1970 article in the British science magazine *Nature*, physicist Pieter van Heerden proposed that a type of holography known as *recognition holography* offers a way of understanding this ability.[*] In recognition holography a holographic image of an object is recorded in the usual manner, save that the laser beam is bounced off a special kind of mirror known as a *focusing mirror* before it is allowed to strike the unexposed film. If a second object, similar but not identical

*Van Heerden, a researcher at the Polaroid Research Laboratories in Cambridge, Massachusetts, actually proposed his own version of a holographic theory of memory in 1963, but his work went relatively unnoticed.

to the first, is bathed in laser light and the light is bounced off the mirror and onto the film after it has been developed, a bright point of light will appear on the film. The brighter and sharper the point of light, the greater the degree of similarity between the first and second objects. If the two objects are completely dissimilar, no point of light will appear. By placing a light-sensitive photocell behind the holographic film, one can actually use the setup as a mechanical recognition system.[7]

A similar technique known as *interference holography* may also explain how we can recognize both the familiar and unfamiliar features of an image such as the face of someone we have not seen for many years. In this technique an object is viewed through a piece of holographic film containing its image. When this is done, any feature of the object that has changed since its image was originally recorded will reflect light differently. An individual looking through the film is instantly aware of both how the object has changed and how it has remained the same. The technique is so sensitive that even the pressure of a finger on a block of granite shows up immediately, and the process has been found to have practical applications in the materials-testing industry.[8]

PHOTOGRAPHIC MEMORY

In 1972, Harvard vision researchers Daniel Pollen and Michael Tractenberg proposed that the holographic brain theory may explain why some people possess photographic memories (also known as *eidetic memories*). Typically, individuals with photographic memories will spend a few moments scanning the scene they wish to memorize. When they want to see the scene again, they "project" a mental image of it, either with their eyes closed or as they gaze at a blank wall or screen. In a study of one such individual, a Harvard art history professor named Elizabeth, Pollen and Tractenberg found that the mental images she projected were so real to her that when she read an image of a page from Goethe's *Faust* her eyes moved as if she were reading a real page.

Noting that the image stored in a fragment of holographic film gets hazier as the fragment gets smaller, Pollen and Tractenberg suggest that perhaps such individuals have more vivid memories because they somehow have access to very large regions of their memory holograms. Conversely, perhaps most of us have memories that are much less vivid because our access is limited to smaller regions of the memory holograms.[9]

THE TRANSFERENCE OF LEARNED SKILLS

Pribram believes the holographic model also sheds light on our ability to transfer learned skills from one part of our body to another. As you sit reading this book, take a moment and trace your first name in the air with your left elbow. You will probably discover that this is a relatively easy thing to do, and yet in all likelihood it is something you have never done before. It may not seem a surprising ability to you, but in the classic view that various areas of the brain (such as the area controlling the movements of the elbow) are "hard-wired," or able to perform tasks *only* after repetitive learning has caused the proper neural connections to become established between brain cells, this is something of a puzzle. Pribram points out that the problem becomes much more tractable if the brain were to convert all of its memories, including memories of learned abilities such as writing, into a language of interfering wave forms. Such a brain would be much more flexible and could shift its stored information around with the same ease that a skilled pianist transposes a song from one musical key to another.

This same flexibility may explain how we are able to recognize a familiar face regardless of the angle from which we are viewing it. Again, once the brain has memorized a face (or any other object or scene) and converted it into a language of wave forms, it can, in a sense, tumble this internal hologram around and examine it from any perspective it wants.

PHANTOM LIMB SENSATIONS AND HOW WE
CONSTRUCT A "WORLD-OUT-THERE"

To most of us it is obvious that our feelings of love, hunger, anger, and so on, are internal realities, and the sound of an orchestra playing, the heat of the sun, the smell of bread baking, and so on, are external realities. But it is not so clear how our brains enable us to distinguish between the two. For example, Pribram points out that when we look at a person, the image of the person is really on the surface of our retinas. Yet we do not perceive the person as being on our retinas. We perceive them as being in the "world-out-there." Similarly, when we stub our toe we experience the pain in our toe. But the pain is not really in our toe. It is actually a neurophysiological process taking place somewhere in our brain. How then is our brain able to take the multitude of neurophysiological processes that manifest as our experience, all of which are internal, and fool us into thinking that some are internal and some are located beyond the confines of our gray matter?

Creating the illusion that things are located where they are not is the quintessential feature of a hologram. As mentioned, if you look at a hologram it seems to have extension in space, but if you pass your hand through it you will discover there is nothing there. Despite what your senses tell you, no instrument will pick up the presence of any abnormal energy or substance where the hologram appears to be hovering. This is because a hologram is a *virtual* image, an image that appears to be where it is not, and possesses no more extension in space than does the three-dimensional image you see of yourself when you look in a mirror. Just as the image in the mirror is located in the silvering on the mirror's back surface, the actual location of a hologram is always in the photographic emulsion on the surface of the film recording it.

Further evidence that the brain is able to fool us into thinking that inner processes are located outside the body comes from the Nobel Prize–winning physiologist Georg von Bekesy. In a series of experiments conducted in the late 1960s Bekesy placed vibrators on the knees of blindfolded test subjects. Then he varied the rates at which the instruments vibrated. By doing so he discovered that he could make his test subjects experience the sensation that a point source of vibration was jumping from one knee to the other. He found that he could even make his subjects feel the point source of vibration in the space *between* their knees. In short, he demonstrated that humans have the ability to seemingly experience sensation in spatial locations where they have absolutely no sense receptors.[10]

Pribram believes that Bekesy's work is compatible with the holographic view and sheds additional light on how interfering wave fronts—or in Bekesy's case, interfering sources of physical vibration—enable the brain to localize some of its experiences beyond the physical boundaries of the body. He feels this process might also explain the phantom limb phenomenon, or the sensation experienced by some amputees that a missing arm or leg is still present. Such individuals often feel eerily realistic cramps, pains, and tinglings in these phantom appendages, but maybe what they are experiencing is the holographic memory of the limb that is still recorded in the interference patterns in their brains.

Experimental Support for the Holographic Brain

For Pribram the many similarities between brains and holograms were tantalizing, but he knew his theory didn't mean anything unless it was backed up by more solid evidence. One researcher who provided such evidence was Indiana University biologist Paul Pietsch. Intriguingly, Pietsch began as an ardent disbeliever in Pribram's theory. He was especially skeptical of Pribram's claim that memories do not possess any specific location in the brain.

To prove Pribram wrong, Pietsch devised a series of experiments, and as the test subjects of his experiments he chose salamanders. In previous studies he had discovered that he could remove the brain of a salamander without killing it, and although it remained in a stupor as long as its brain was missing, its behavior completely returned to normal as soon as its brain was restored.

Pietsch reasoned that if a salamander's feeding behavior is not confined to any specific location in the brain, then it should not matter how its brain is positioned in its head. If it did matter, Pribram's theory would be disproven. He then flip-flopped the left and right hemispheres of a salamander's brain, but to his dismay, as soon as it recovered, the salamander quickly resumed normal feeding.

He took another salamander and turned its brain upside down. When it recovered it, too, fed normally. Growing increasingly frustrated, he decided to resort to more drastic measures. In a series of over 700 operations he sliced, flipped, shuffled, subtracted, and even minced the brains of his hapless subjects, but always when he replaced what was left of their brains, their behavior returned to normal.[11]

These findings and others turned Pietsch into a believer and attracted enough attention that his research became the subject of a segment on the television show *60 Minutes*. He writes about this experience as well as giving detailed accounts of his experiments in his insightful book *Shufflebrain*.

The Mathematical Language of the Hologram

While the theories that enabled the development of the hologram were first formulated in 1947 by Dennis Gabor (who later won a Nobel Prize for his efforts), in the late 1960s and early 1970s Pribram's theory received even more persuasive experimental support. When Gabor first conceived the idea of holography he wasn't thinking about lasers. His goal was to improve the electron microscope, then a primitive and imperfect device. His approach was a mathematical one, and the mathematics he used was a type of calculus invented by an eighteenth-century Frenchman named Jean B. J. Fourier.

Roughly speaking what Fourier developed was a mathematical way of converting any pattern, no matter how complex, into a language of simple waves. He also showed how these wave forms could be converted back into the original pattern. In other words, just as a television camera converts an image into electromagnetic frequencies and a television set converts those frequencies back into the original image, Fourier showed how a similar process could be achieved mathematically. The equations he developed to convert images into wave forms and back again are known as *Fourier transforms*.

Fourier transforms enabled Gabor to convert a picture of an object into the blur of interference patterns on a piece of holographic film. They also enabled him to devise a way of converting those interference patterns back into an image of the original object. In fact the special whole in every part of a hologram is one of the by-products that occurs when an image or pattern is translated into the Fourier language of wave forms.

Throughout the late 1960s and early 1970s various researchers contacted Pribram and told him they had uncovered evidence that the visual system worked as a kind of frequency analyzer. Since frequency is a measure of the number of oscillations a wave undergoes per second, this strongly suggested that the brain might be functioning as a hologram does.

But it wasn't until 1979 that Berkeley neurophysiologists Russell and Karen DeValois made the discovery that settled the matter. Research in the 1960s had shown that each brain cell in the visual cortex is geared to respond to a different pattern—some brain cells fire when the eyes see a horizontal line, others fire when the eyes see a vertical line, and so on. As a result, many researchers concluded that the brain takes input from these highly specialized cells called feature detectors, and somehow fits them together to provide us with our visual perceptions of the world.

Despite the popularity of this view, the DeValoises felt it was only a partial truth. To test their assumption they used Fourier's equations to convert plaid and checkerboard patterns into simple wave forms. Then they tested to see how the brain cells in the visual cortex responded to these new wave-form images. What they found was that the brain cells responded not to the original patterns, but to the Fourier translations of the patterns. Only one conclusion could be drawn. The brain was using Fourier mathematics—the same mathematics holography employed—to convert visual images into the Fourier language of wave forms.[12]

The DeValoises' discovery was subsequently confirmed by numerous other laboratories around the world, and although it did not provide absolute proof the brain was a hologram, it supplied enough evidence to convince Pribram his theory was correct. Spurred on by the idea that the visual cortex was responding not to patterns but to the frequencies of various wave forms, he began to reassess the role frequency played in the other senses.

It didn't take long for him to realize that the importance of this role had perhaps been overlooked by twentieth-century scientists. Over a century before the DeValoises' discovery, the German physiologist and physicist Hermann von Helmholtz had shown that the ear was a frequency analyzer. More recent research revealed that our sense of smell seems to be based on what are called osmic frequencies. Bekesy's work had clearly demonstrated that our skin is sensitive to frequencies of vibration, and he even produced some evidence that taste may involve frequency analysis. Interestingly, Bekesy also discovered that the mathematical equations that enabled him to predict how his subjects would respond to various frequencies of vibration were also of the Fourier genre.

The Dancer as Wave Form

But perhaps the most startling finding Pribram uncovered was Russian scientist Nikolai Bernstein's discovery that even our physical movements may be encoded in our brains in a language of Fourier wave forms. In the 1930s Bernstein dressed people in black leotards

FIGURE 6. Russian researcher Nikolai Bernstein painted white dots on dancers and filmed them dancing against a black background. When he converted their movements into a language of wave forms, he discovered they could be analyzed using Fourier mathematics, the same mathematics Gabor used to invent the hologram.

and painted white dots on their elbows, knees, and other joints. Then he placed them against black backgrounds and took movies of them doing various physical activities such as dancing, walking, jumping, hammering, and typing.

When he developed the film, only the white dots appeared, moving up and down and across the screen in various complex and flowing movements (see fig. 6). To quantify his findings he Fourier-analyzed the various lines the dots traced out and converted them into a language of wave forms. To his surprise, he discovered the wave forms contained hidden patterns that allowed him to predict his subjects' next movement to within a fraction of an inch.

When Pribram encountered Bernstein's work he immediately recognized its implications. Maybe the reason hidden patterns surfaced after Bernstein Fourier-analyzed his subject's movements was because that was how movements are stored in the brain. This was an exciting possibility, for if the brain analyzed movements by breaking them down into their frequency components, it explained the rapidity with which we learn many complex physical tasks. For instance, we do not learn to ride a bicycle by painstakingly memorizing every tiny feature of the process. We learn by grasping the whole flowing movement. The fluid wholeness that typifies how we learn so many physical activities is difficult to explain if our brains are storing information in a bit-by-bit manner. But it becomes much easier to understand if the brain is Fourier-analyzing such tasks and absorbing them as a whole.

The Reaction of the Scientific Community

Despite such evidence, Pribram's holographic model remains extremely controversial. Part of the problem is that there are many popular theories of how the brain works and there is evidence to support them all. Some researchers believe the distributed nature of memory can be explained by the ebb and flow of various brain chemicals. Others hold that electrical fluctuations among large groups of neurons can account for memory and learning. Each school of thought has its ardent supporters, and it is probably safe to say that most scientists remain unpersuaded by Pribram's arguments. For example, neuropsychologist Frank Wood of the Bowman Gray School of Medicine in Winston-Salem, North Carolina, feels that "there are precious few experimental findings for which holography is the necessary, or even preferable, explanation."[13] Pribram is puzzled by statements such as Wood's and counters by noting that he currently has a book in press with well over 500 references to such data.

Other researchers agree with Pribram. Dr. Larry Dossey, former chief of staff at Medical City Dallas Hospital, admits that Pribram's theory challenges many long-held assumptions about the brain, but points out that "many specialists in brain function are attracted to the idea, if for no other reason than the glaring inadequacies of the present orthodox views."[14]

Neurologist Richard Restak, author of the PBS series *The Brain*, shares Dossey's opinion. He notes that in spite of overwhelming evidence that human abilities are holistically dispersed throughout the brain, most researchers continue to cling to the idea that function can be located in the brain in the same way that cities can be located on a map. Restak believes that theories based on this premise are not only "oversimplistic," but actually function as "conceptual straitjackets" that keep us from recognizing the brain's true complexities.[15] He feels that "a hologram is not only possible but, at this moment, represents probably our best 'model' for brain functioning."[16]

Pribram Encounters Bohm

As for Pribram, by the 1970s enough evidence had accumulated to convince him his theory was correct. In addition, he had taken his ideas into the laboratory and discovered that single neurons in the motor cortex respond selectively to a limited bandwidth of frequencies, a finding that further supported his conclusions. The question that began to bother him was, If the picture of reality in our brains is not a picture at all but a hologram, what is it a hologram of? The dilemma posed by this question is analogous to taking a Polaroid picture of a group of people sitting around a table and, after the picture develops, finding that, instead of people, there are only blurry clouds of interference patterns positioned around the table. In both cases one could rightfully ask, Which is the true reality, the seemingly objective world experienced by the observer/photographer or the blur of interference patterns recorded by the camera/brain?

Pribram realized that if the holographic brain model was taken to its logical conclusions, it opened the door on the possibility that objective reality—the world of coffee cups, mountain vistas, elm trees, and table lamps—might not even exist, or at least not exist in the way we believe it exists. Was it possible, he wondered, that what the mystics had been saying for centuries was true, reality was *maya*, an illusion, and what was out there was really a vast, resonating symphony of wave forms, a "frequency domain" that was transformed into the world as we know it only *after* it entered our senses?

Realizing that the solution he was seeking might lie outside the province of his own field, he went to his physicist son for advice. His son recommended he look into the work of a physicist named David Bohm. When Pribram did he was electrified. He not only found the answer to his question, but also discovered that according to Bohm, the entire universe was a hologram.

2

The Cosmos as Hologram

One can't help but be astonished at the degree to which
[Bohm] has been able to break out of the tight molds of scientific conditioning and stand alone with a completely new and
literally vast idea, one which has both internal consistency and
the logical power to explain widely diverging phenomena of
physical experience from an entirely unexpected point of view.
. . . It is a theory which is so intuitively satisfying that many
people have felt that if the universe is not the way Bohm
describes it, it ought to be.

—John P. Briggs and F. David Peat
Looking Glass Universe

The path that led Bohm to the conviction that the universe is struc-
tured like a hologram began at the very edge of matter, in the world
of subatomic particles. His interest in science and the way things work
blossomed early. As a young boy growing up in Wilkes-Barre, Pennsylvania, he invented a dripless tea kettle, and his father, a successful
businessman, urged him to try to turn a profit on the idea. But after
learning that the first step in such a venture was to conduct a door-to-
door survey to test-market his invention, Bohm's interest in business
waned.[1]

His interest in science did not, however, and his prodigious curiosity
forced him to look for new heights to conquer. He found the most
challenging height of all in the 1930s when he attended Pennsylvania

State College, for it was there that he first became fascinated by quantum physics.

It is an easy fascination to understand. The strange new land that physicists had found lurking in the heart of the atom contained things more wondrous than anything Cortés or Marco Polo ever encountered. What made this new world so intriguing was that everything about it appeared to be so contrary to common sense. It seemed more like a land ruled by sorcery than an extension of the natural world, an Alice-in-Wonderland realm in which mystifying forces were the norm and everything logical had been turned on its ear.

One startling discovery made by quantum physicists was that if you break matter into smaller and smaller pieces you eventually reach a point where those pieces—electrons, protons, and so on—no longer possess the traits of objects. For example, most of us tend to think of an electron as a tiny sphere or a BB whizzing around, but nothing could be further from the truth. Although an electron can sometimes behave as if it were a compact little particle, physicists have found that *it literally possesses no dimension.* This is difficult for most of us to imagine because everything at our own level of existence possesses dimension. And yet if you try to measure the width of an electron, you will discover it's an impossible task. An electron is simply not an object as we know it.

Another discovery physicists made is that an electron can manifest as either a particle or a wave. If you shoot an electron at the screen of a television that's been turned off, a tiny point of light will appear when it strikes the phosphorescent chemicals that coat the glass. The single point of impact the electron leaves on the screen clearly reveals the particlelike side of its nature.

But this is not the only form the electron can assume. It can also dissolve into a blurry cloud of energy and behave as if it were a wave spread out over space. When an electron manifests as a wave it can do things no particle can. If it is fired at a barrier in which two slits have been cut, it can go through both slits simultaneously. When wavelike electrons collide with each other they even create interference patterns. The electron, like some shapeshifter out of folklore, can manifest as either a particle or a wave.

This chameleonlike ability is common to all subatomic particles. It is also common to all things once thought to manifest exclusively as waves. Light, gamma rays, radio waves, X rays—all can change from waves to particles and back again. Today physicists believe that sub-

atomic phenomena should not be classified solely as either waves or
particles, but as a single category of somethings that are always
somehow both. These somethings are called *quanta*, and physicists
believe they are the basic stuff from which the entire universe is
made.*

Perhaps most astonishing of all is that there is compelling evidence
that *the only time quanta ever manifest as particles is when we are
looking at them.* For instance, when an electron isn't being looked at,
experimental findings suggest that it is always a wave. Physicists are
able to draw this conclusion because they have devised clever strate-
gies for deducing how an electron behaves when it is not being observed (it should be noted that this is only one interpretation of the
evidence and is not the conclusion of all physicists; as we will see,
Bohm himself has a different interpretation).

Once again this seems more like magic than the kind of behavior we
are accustomed to expect from the natural world. Imagine owning a
bowling ball that was only a bowling ball when you looked at it. If you
sprinkled talcum powder all over a bowling lane and rolled such a
"quantum" bowling ball toward the pins, it would trace a single line
through the talcum powder while you were watching it. But if you
blinked while it was in transit, you would find that for the second or
two you were not looking at it the bowling ball stopped tracing a line
and instead left a broad wavy strip, like the undulating swath of a
desert snake as it moves sideways over the sand (see fig. 7).

Such a situation is comparable to the one quantum physicists encountered when they first uncovered evidence that quanta coalesce
into particles only when they are being observed. Physicist Nick Herbert, a supporter of this interpretation, says this has sometimes
caused him to imagine that behind his back the world is always "a
radically ambiguous and ceaselessly flowing quantum soup." But
whenever he turns around and tries to see the soup, his glance instantly freezes it and turns it back into ordinary reality. He believes
this makes us all a little like Midas, the legendary king who never
knew the feel of silk or the caress of a human hand because everything
he touched turned to gold. "Likewise humans can never experience the
true texture of quantum reality," says Herbert, "because everything
we touch turns to matter."[2]

Quanta is the plural of *quantum*. One electron is a quantum. Several electrons are a
group of quanta. The word *quantum* is also synonymous with *wave particle*, a term that
is also used to refer to something that possesses both particle and wave aspects.

Bohm and Interconnectedness

An aspect of quantum reality that Bohm found especially interesting was the strange state of interconnectedness that seemed to exist between apparently unrelated subatomic events. What was equally perplexing was that most physicists tended to attach little importance to the phenomenon. In fact, so little was made of it that one of the most famous examples of interconnectedness lay hidden in one of quantum physics's basic assumptions for a number of years before anyone noticed it was there.

That assumption was made by one of the founding fathers of quantum physics, the Danish physicist Niels Bohr. Bohr pointed out that if subatomic particles only come into existence in the presence of an observer, then it is also meaningless to speak of a particle's properties and characteristics as existing before they are observed. This was disturbing to many physicists, for much of science was based on discovering the properties of phenomena. But if the act of observation actually helped create such properties, what did that imply about the future of science?

One physicist who was troubled by Bohr's assertions was Einstein. Despite the role Einstein had played in the founding of quantum theory, he was not at all happy with the course the fledgling science

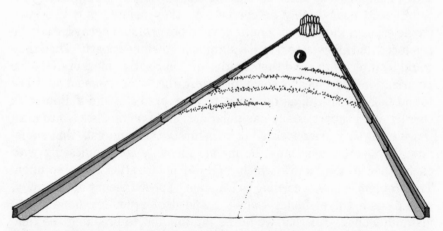

FIGURE 7. Physicists have found compelling evidence that the only time electrons and other "quanta" manifest as particles is when we are looking at them. At all other times they behave as waves. This is as strange as owning a bowling ball that traces a single line down the lane while you are watching it, but leaves a wave pattern every time you blink your eyes.

had taken. He found Bohr's conclusion that a particle's properties don't exist until they are observed particularly objectionable because, when combined with another of quantum physics's findings, it implied that subatomic particles were interconnected in a way Einstein simply didn't believe was possible.

That finding was the discovery that some subatomic processes result in the creation of a pair of particles with identical or closely related properties. Consider an extremely unstable atom physicists call positronium. The positronium atom is composed of an electron and a positron (a positron is an electron with a positive charge). Because a positron is the electron's antiparticle opposite, the two eventually annihilate each other and decay into two quanta of light or "photons" traveling in opposite directions (the capacity to shapeshift from one kind of particle to another is just another of a quantum's abilities). According to quantum physics no matter how far apart the photons travel, when they are measured they will always be found to have identical angles of *polarization*. (Polarization is the spatial orientation of the photon's wavelike aspect as it travels away from its point of origin.)

In 1935 Einstein and his colleagues Boris Podolsky and Nathan Rosen published a now famous paper entitled "Can Quantum-Mechanical Description of Physical Reality Be Considered Complete?" In it they explained why the existence of such twin particles proved that Bohr could not possibly be correct. As they pointed out, two such particles, say, the photons emitted when positronium decays, could be produced and allowed to travel a significant distance apart.* Then they could be intercepted and their angles of polarization measured. If the polarizations are measured at precisely the same moment and are found to be identical, as quantum physics predicts, and if Bohr was correct and properties such as polarization do not coalesce into existence until they are observed or measured, this suggests that somehow the two photons must be instantaneously communicating with each other so they know which angle of polarization to agree upon. The problem is that according to Einstein's special theory of relativity, nothing can travel faster than the speed of light, let alone travel instantaneously, for that would be tantamount to breaking the time

---

*Positronium decay is not the subatomic process Einstein and his colleagues employed in their thought experiment, but is used here because it is easy to visualize.

barrier and would open the door on all kinds of unacceptable paradoxes. Einstein and his colleagues were convinced that no "reasonable definition" of reality would permit such faster-than-light interconnections to exist, and therefore Bohr had to be wrong.[3] Their argument is now known as the Einstein-Podolsky-Rosen paradox, or EPR paradox for short.

Bohr remained unperturbed by Einstein's argument. Rather than believing that some kind of faster-than-light communication was taking place, he offered another explanation. If subatomic particles do not exist until they are observed, then one could no longer think of them as independent "things." Thus Einstein was basing his argument on an error when he viewed twin particles as separate. They were part of an indivisible system, and it was meaningless to think of them otherwise.

In time most physicists sided with Bohr and became content that his interpretation was correct. One factor that contributed to Bohr's triumph was that quantum physics had proved so spectacularly successful in predicting phenomena, few physicists were willing even to consider the possibility that it might be faulty in some way. In addition, when Einstein and his colleagues first made their proposal about twin particles, technical and other reasons prevented such an experiment from actually being performed. This made it even easier to put out of mind. This was curious, for although Bohr had designed his argument to counter Einstein's attack on quantum theory, as we will see, Bohr's view that subatomic systems are indivisible has equally profound implications for the nature of reality. Ironically, these implications were also ignored, and once again the potential importance of interconnectedness was swept under the carpet.

A Living Sea of Electrons

During his early years as a physicist Bohm also accepted Bohr's position, but he remained puzzled by the lack of interest Bohr and his followers displayed toward interconnectedness. After graduating from Pennsylvania State College, he attended the University of California at Berkeley, and before receiving his doctorate there in 1943, he worked at the Lawrence Berkeley Radiation Laboratory. There he encountered another striking example of quantum interconnectedness.

At the Berkeley Radiation Laboratory Bohm began what was to become his landmark work on plasmas. A plasma is a gas containing a high density of electrons and positive ions, atoms that have a positive charge. To his amazement he found that once they were in a plasma, electrons stopped behaving like individuals and started behaving as if they were part of a larger and interconnected whole. Although their individual movements appeared random, vast numbers of electrons were able to produce effects that were surprisingly well-organized. Like some amoeboid creature, the plasma constantly regenerated itself and enclosed all impurities in a wall in the same way that a biological organism might encase a foreign substance in a cyst.[4] So struck was Bohm by these organic qualities that he later remarked he'd frequently had the impression the electron sea was "alive."[5]

In 1947 Bohm accepted an assistant professorship at Princeton University, an indication of how highly he was regarded, and there he extended his Berkeley research to the study of electrons in metals. Once again he found that the seemingly haphazard movements of individual electrons managed to produce highly organized overall effects. Like the plasmas he had studied at Berkeley, these were no longer situations involving two particles, each behaving as if it knew what the other was doing, but entire oceans of particles, each behaving as if it knew what untold trillions of others were doing. Bohm called such collective movements of electrons *plasmons*, and their discovery established his reputation as a physicist.

Bohm's Disillusionment

Both his sense of the importance of interconnectedness as well as his growing dissatisfaction with several of the other prevailing views in physics caused Bohm to become increasingly troubled by Bohr's interpretation of quantum theory. After three years of teaching the subject at Princeton he decided to improve his understanding by writing a textbook. When he finished he found he still wasn't comfortable with what quantum physics was saying and sent copies of the book to both Bohr and Einstein to ask for their opinions. He got no answer from Bohr, but Einstein contacted him and said that since they were both at Princeton they should meet and discuss the book. In the first of what was to turn into a six-month series of spirited conversations, Einstein enthusiastically told Bohm that he had never seen quantum theory presented so clearly. Nonetheless, he admitted he was still every bit as dissatisfied with the theory as was Bohm.

During their conversations the two men discovered they each had nothing but admiration for the theory's ability to predict phenomena. What bothered them was that it provided no real way of conceiving of the basic structure of the world. Bohr and his followers also claimed that quantum theory was complete and it was not possible to arrive at any clearer understanding of what was going on in the quantum realm. This was the same as saying there was no deeper reality beyond the subatomic landscape, no further answers to be found, and this, too, grated on both Bohm and Einstein's philosophical sensibilities. Over the course of their meetings they discussed many other things, but these points in particular gained new prominence in Bohm's thoughts. Inspired by his interactions with Einstein, he accepted the validity of his misgivings about quantum physics and decided there had to be an alternative view. When his textbook *Quantum Theory* was published in 1951 it was hailed as a classic, but it was a classic about a subject to which Bohm no longer gave his full allegiance. His mind, ever active and always looking for deeper explanations, was already searching for a better way of describing reality.

A New Kind of Field and the Bullet That Killed Lincoln

After his talks with Einstein, Bohm tried to find a workable alternative to Bohr's interpretation. He began by assuming that particles such as electrons *do* exist in the absence of observers. He also assumed that there was a deeper reality beneath Bohr's inviolable wall, a subquantum level that still awaited discovery by science. Building on these premises he discovered that simply by proposing the existence of a new kind of field on this subquantum level he was able to explain the findings of quantum physics as well as Bohr could. Bohm called his proposed new field the *quantum potential* and theorized that, like gravity, it pervaded all of space. However, unlike gravitational fields, magnetic fields, and so on, its influence did not diminish with distance. Its effects were subtle, but it was equally powerful everywhere. Bohm published his alternative interpretation of quantum theory in 1952.

Reaction to his new approach was mainly negative. Some physicists were so convinced such alternatives were impossible that they dismissed his ideas out of hand. Others launched passionate attacks against his reasoning. In the end virtually all such arguments were based primarily on philosophical differences, but it did not matter. Bohr's point of view had become so entrenched in physics that Bohm's alternative was looked upon as little more than heresy.

Despite the harshness of these attacks Bohm remained unswerving in his conviction that there was more to reality than Bohr's view allowed. He also felt that science was much too limited in its outlook when it came to assessing new ideas such as his own, and in a 1957 book entitled *Causality and Chance in Modern Physics*, he examined several of the philosophical suppositions responsible for this attitude. One was the widely held assumption that it was possible for any single theory, such as quantum theory, to *be* complete. Bohm criticized this assumption by pointing out that nature may be infinite. Because it would not be possible for any theory to completely explain something that is infinite, Bohm suggested that open scientific inquiry might be better served if researchers refrained from making this assumption.

In the book he argued that the way science viewed causality was also much too limited. Most effects were thought of as having only one or several causes. However, Bohm felt that an effect could have an infinite number of causes. For example, if you asked someone what caused Abraham Lincoln's death, they might answer that it was the bullet in John Wilkes Booth's gun. But a complete list of all the causes that contributed to Lincoln's death would have to include all of the events that led to the development of the gun, all of the factors that caused Booth to want to kill Lincoln, all of the steps in the evolution of the human race that allowed for the development of a hand capable of holding a gun, and so on, and so on. Bohm conceded that most of the time one could ignore the vast cascade of causes that had led to any given effect, but he still felt it was important for scientists to remember that no single cause-and-effect relationship was ever really separate from the universe as a whole.

If You Want to Know Where You Are, Ask the Nonlocals

During this same period of his life Bohm also continued to refine his alternative approach to quantum physics. As he looked more carefully into the meaning of the quantum potential he discovered it had a number of features that implied an even more radical departure from orthodox thinking. One was the importance of wholeness. Classical science had always viewed the state of a system as a whole as merely the result of the interaction of its parts. However, the quantum potential stood this view on its ear and indicated that the behavior of the parts was actually organized by the whole. This not only took Bohr's assertion that subatomic particles are not independent "things," but are part of an indivisible system one step further, but even suggested that wholeness was in some ways the more primary reality.

It also explained how electrons in plasmas (and other specialized states such as superconductivity) could behave like interconnected wholes. As Bohm states, such "electrons are not scattered because, through the action of the quantum potential, the whole system is undergoing a co-ordinated movement more like a ballet dance than like a crowd of unorganized people." Once again he notes that "such quantum wholeness of activity is closer to the organized unity of functioning of the parts of a living being than it is to the kind of unity that is obtained by putting together the parts of a machine."[6]

An even more surprising feature of the quantum potential was its implications for the nature of location. At the level of our everyday lives things have very specific locations, but Bohm's interpretation of quantum physics indicated that at the subquantum level, the level in which the quantum potential operated, location ceased to exist. All points in space became equal to all other points in space, and it was meaningless to speak of anything as being separate from anything else. Physicists call this property "nonlocality."

The nonlocal aspect of the quantum potential enabled Bohm to explain the connection between twin particles without violating special relativity's ban against anything traveling faster than the speed of light. To illustrate how, he offers the following analogy: Imagine a fish swimming in an aquarium. Imagine also that you have never seen a fish or an aquarium before and your only knowledge about them comes from two television cameras, one directed at the aquarium's front and the other at its side. When you look at the two television monitors you might mistakenly assume that the fish on the screens are separate entities. After all, because the cameras are set at different angles, each of the images will be slightly different. But as you continue to watch you will eventually realize there is a relationship between the two fish. When one turns, the other makes a slightly different but corresponding turn. When one faces the front, the other faces the side, and so on. If you are unaware of the full scope of the situation, you might wrongly conclude that the fish are instantaneously communicating with one another, but this is not the case. No communication is taking place because at a deeper level of reality, the reality of the aquarium, the two fish are actually one and the same. This, says Bohm, is precisely what is going on between particles such as the two photons emitted when a positronium atom decays (see fig. 8).

Indeed, because the quantum potential permeates all of space, all

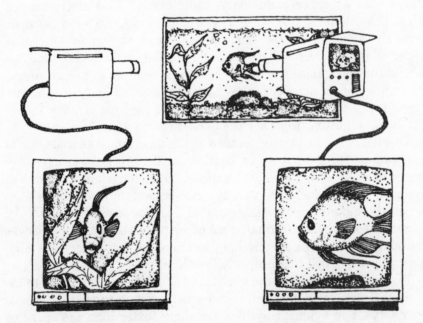

FIGURE 8. Bohm believes subatomic particles are connected in the same way as the images of the fish on the two television monitors. Although particles such as electrons appear to be separate from one another, on a deeper level of reality—a level analogous to the aquarium—they are actually just different aspects of a deeper cosmic unity.

particles are nonlocally interconnected. More and more the picture of reality Bohm was developing was not one in which subatomic particles were separate from one another and moving through the void of space, but one in which all things were part of an unbroken web and embedded in a space that was as real and rich with process as the matter that moved through it.

Bohm's ideas still left most physicists unpersuaded, but did stir the interest of a few. One of these was John Stewart Bell, a theoretical physicist at CERN, a center for peaceful atomic research near Geneva, Switzerland. Like Bohm, Bell had also become discontented with quantum theory and felt there must be some alternative. As he later said, "Then in 1952 I saw Bohm's paper. His idea was to complete quantum mechanics by saying there are certain variables in addition to those which everybody knew about. That impressed me very much."[7]

Bell also realized that Bohm's theory implied the existence of nonlocality and wondered if there was any way of experimentally verifying its existence. The question remained in the back of his mind for years until a sabbatical in 1964 provided him with the freedom to focus his full attention on the matter. Then he quickly came up with an elegant mathematical proof that revealed how such an experiment could be performed. The only problem was that it required a level of technological precision that was not yet available. To be certain that particles, such as those in the EPR paradox, were not using some normal means of communication, the basic operations of the experiment had to be performed in such an infinitesimally brief instant that there wouldn't even be enough time for a ray of light to cross the distance separating the two particles. This meant that the instruments used in the experiment had to perform all of the necessary operations within a few thousand-millionths of a second.

Enter the Hologram

By the late 1950s Bohm had already had his run-in with McCarthyism and had become a research fellow at Bristol University, England. There, along with a young research student named Yakir Aharonov, he discovered another important example of nonlocal interconnectedness. Bohm and Aharonov found that under the right circumstances an electron is able to "feel" the presence of a magnetic field that is in a region where there is zero probability of finding the electron. This phenomenon is now known as the Aharonov-Bohm effect, and when the two men first published their discovery, many physicists did not believe such an effect was possible. Even today there is enough residual skepticism that, despite confirmation of the effect in numerous experiments, occasionally papers still appear arguing that it doesn't exist.

As always, Bohm stoically accepted his continuing role as the voice in the crowd that bravely notes the emperor has no clothes. In an interview conducted some years later he offered a simple summation of the philosophy underlying his courage: "In the long run it is far more dangerous to adhere to illusion than to face what the actual fact is."[8]

Nevertheless, the limited response to his ideas about wholeness and nonlocality and his own inability to see how to proceed further caused him to focus his attention in other directions. In the 1960s this led him to take a closer look at *order*. Classical science generally divides things into two categories: those that possess order in the arrangement of their parts and those whose parts are disordered, or random, in arrangement. Snowflakes, computers, and living things are all ordered. The pattern a handful of spilled coffee beans makes on the floor, the debris left by an explosion, and a series of numbers generated by a roulette wheel are all disordered.

As Bohm delved more deeply into the matter he realized there were also different degrees of order. Some things were much more ordered than other things, and this implied that there was, perhaps, no end to the hierarchies of order that existed in the universe. From this it occurred to Bohm that maybe things that we perceive as disordered aren't disordered at all. Perhaps their order is of such an "indefinitely high degree" that they only appear to us as random (interestingly, mathematicians are unable to prove randomness, and although some sequences of numbers are categorized as random, these are only educated guesses).

While immersed in these thoughts, Bohm saw a device on a BBC television program that helped him develop his ideas even further. The device was a specially designed jar containing a large rotating cylinder. The narrow space between the cylinder and the jar was filled with glycerine—a thick, clear liquid—and floating motionlessly in the glycerine was a drop of ink. What interested Bohm was that when the handle on the cylinder was turned, the drop of ink spread out through the syrupy glycerine and seemed to disappear. But as soon as the handle was turned back in the opposite direction, the faint tracing of ink slowly collapsed upon itself and once again formed a droplet (see fig. 9).

Bohm writes, "This immediately struck me as very relevant to the question of order, since, when the ink drop was spread out, it still had a 'hidden' (i.e., nonmanifest) order that was revealed when it was reconstituted. On the other hand, in our usual language, we would say that the ink was in a state of 'disorder' when it was diffused through the glycerine. This led me to see that new notions of order must be involved here."[9]

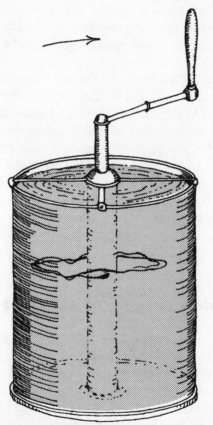

FIGURE 9. When a drop of ink is placed in a jar full of glycerine and a cylinder inside the jar is turned, the drop appears to spread out and disappear. But when the cylinder is turned in the opposite direction, the drop comes back together. Bohm uses this phenomenon as an example of how order can be either manifest (explicit) or hidden (implicit).

This discovery excited Bohm greatly, for it provided him with a new way of looking at many of the problems he had been contemplating. Soon after coming across the ink-in-glycerine device he encountered an even better metaphor for understanding order, one that enabled him not only to bring together all the various strands of his years of thinking, but did so with such force and explanatory power it seemed almost tailor-made for the purpose. That metaphor was the hologram.

As soon as Bohm began to reflect on the hologram he saw that it *too* provided a new way of understanding order. Like the ink drop in its dispersed state, the interference patterns recorded on a piece of holographic film also appear disordered to the naked eye. Both possess orders that are hidden or *enfolded* in much the same way that the order in a plasma is enfolded in the seemingly random behavior of each of its electrons. But this was not the only insight the hologram provided.

The more Bohm thought about it the more convinced he became that the universe actually employed holographic principles in its operations, *was itself a kind of giant, flowing hologram,* and this realization allowed him to crystallize all of his various insights into a sweeping and cohesive whole. He published his first papers on his holographic view of the universe in the early 1970s, and in 1980 he presented a mature distillation of his thoughts in a book entitled *Wholeness and the Implicate Order.* In it he did more than just link his myriad ideas together. He transfigured them into a new way of looking at reality that was as breathtaking as it was radical.

Enfolded Orders and Unfolded Realities

One of Bohm's most startling assertions is that the tangible reality of our everyday lives is really a kind of illusion, like a holographic image. Underlying it is a deeper order of existence, a vast and more primary level of reality that gives birth to all the objects and appearances of our physical world in much the same way that a piece of holographic film gives birth to a hologram. Bohm calls this deeper level of reality the *implicate* (which means "enfolded") order, and he refers to our own level of existence as the *explicate,* or unfolded, order.

He uses these terms because he sees the manifestation of all forms in the universe as the result of countless enfoldings and unfoldings between these two orders. For example, Bohm believes an electron is not one thing but a totality or ensemble enfolded throughout the whole of space. When an instrument detects the presence of a single electron it is simply because one aspect of the electron's ensemble has unfolded, similar to the way an ink drop unfolds out of the glycerine, at that particular location. When an electron appears to be moving it is due to a continuous series of such unfoldments and enfoldments.

Put another way, electrons and all other particles are no more substantive or permanent than the form a geyser of water takes as it gushes out of a fountain. They are sustained by a constant influx from the implicate order, and when a particle appears to be destroyed, it is not lost. It has merely enfolded back into the deeper order from which it sprang. A piece of holographic film and the image it generates are also an example of an implicate and explicate order. The film is an implicate order because the image encoded in its interference patterns is a hidden totality enfolded throughout the whole. The hologram projected from the film is an explicate order because it represents the unfolded and perceptible version of the image.

The constant and flowing exchange between the two orders explains how particles, such as the electron in the positronium atom, can shape-shift from one kind of particle to another. Such shiftings can be viewed as one particle, say an electron, enfolding back into the implicate order while another, a photon, unfolds and takes its place. It also explains how a quantum can manifest as either a particle or a wave. According to Bohm, both aspects are always enfolded in a quantum's ensemble, but the way an observer interacts with the ensemble determines which aspect unfolds and which remains hidden. As such, the role an observer plays in determining the form a quantum takes may be no more mysterious than the fact that the way a jeweler manipulates a gem determines which of its facets become visible and which do not. Because the term *hologram* usually refers to an image that is static and does not convey the dynamic and ever active nature of the incalculable enfoldings and unfoldings that moment by moment create our universe, Bohm prefers to describe the universe not as a hologram, but as a "holomovement."

The existence of a deeper and holographically organized order also explains why reality becomes nonlocal at the subquantum level. As we have seen, when something is organized holographically, all semblance of location breaks down. Saying that every part of a piece of holographic film contains all the information possessed by the whole is really just another way of saying that the information is distributed nonlocally. Hence, if the universe is organized according to holographic principles, it, too, would be expected to have nonlocal properties.

The Undivided Wholeness of All Things

Most mind-boggling of all are Bohm's fully developed ideas about wholeness. Because everything in the cosmos is made out of the seamless holographic fabric of the implicate order, he believes it is as meaningless to view the universe as composed of "parts," as it is to view the different geysers in a fountain as separate from the water out of which they flow. An electron is not an "elementary particle." It is just a name given to a certain aspect of the holomovement. Dividing reality up into parts and then naming those parts is always arbitrary, a product of convention, because subatomic particles, and everything else in the universe, are no more separate from one another than different patterns in an ornate carpet.

This is a profound suggestion. In his general theory of relativity Einstein astounded the world when he said that space and time are not separate entities, but are smoothly linked and part of a larger whole he called the space-time continuum. Bohm takes this idea a giant step further. He says that *everything* in the universe is part of a continuum. Despite the apparent separateness of things at the explicate level, everything is a seamless extension of everything else, and ultimately even the implicate and explicate orders blend into each other.

Take a moment to consider this. Look at your hand. Now look at the light streaming from the lamp beside you. And at the dog resting at your feet. You are not merely made of the same things. *You are the same thing.* One thing. Unbroken. One enormous something that has extended its uncountable arms and appendages into all the apparent objects, atoms, restless oceans, and twinkling stars in the cosmos.

Bohm cautions that this does not mean the universe is a giant undifferentiated mass. Things can be part of an undivided whole and still possess their own unique qualities. To illustrate what he means he points to the little eddies and whirlpools that often form in a river. At a glance such eddies appear to be separate things and possess many individual characteristics such as size, rate, and direction of rotation, et cetera. But careful scrutiny reveals that it is impossible to determine where any given whirlpool ends and the river begins. Thus, Bohm is not suggesting that the differences between "things" is meaningless. He merely wants us to be aware constantly that dividing various aspects of the holomovement into "things" is always an abstraction, a way of making those aspects stand out in our perception by our way of thinking. In attempts to correct this, instead of calling different aspects of the holomovement "things," he prefers to call them "relatively independent subtotalities."[10]

Indeed, Bohm believes that our almost universal tendency to fragment the world and ignore the dynamic interconnectedness of all things is responsible for many of our problems, not only in science but in our lives and our society as well. For instance, we believe we can extract the valuable parts of the earth without affecting the whole. We believe it is possible to treat parts of our body and not be concerned with the whole. We believe we can deal with various problems in our society, such as crime, poverty, and drug addiction, without addressing the problems in our society as a whole, and so on. In his writings Bohm argues passionately that our current way of fragmenting the world into parts not only doesn't work, but may even lead to our extinction.

Consciousness as a More Subtle Form of Matter

In addition to explaining why quantum physicists find so many examples of interconnectedness when they plumb the depths of matter, Bohm's holographic universe explains many other puzzles. One is the effect consciousness seems to have on the subatomic world. As we have seen, Bohm rejects the idea that particles don't exist until they are observed. But he is not in principle against trying to bring consciousness and physics together. He simply feels that most physicists go about it the wrong way, by once again trying to fragment reality and saying that one separate thing, consciousness, interacts with another separate thing, a subatomic particle.

Because all such things are aspects of the holomovement, he feels it has no meaning to speak of consciousness and matter as interacting. In a sense, the observer *is* the observed. The observer is also the measuring device, the experimental results, the laboratory, and the breeze that blows outside the laboratory. In fact, Bohm believes that consciousness is a more subtle form of matter, and the basis for any relationship between the two lies not in our own level of reality, but deep in the implicate order. Consciousness is present in various degrees of enfoldment and unfoldment in all matter, which is perhaps why plasmas possess some of the traits of living things. As Bohm puts it, "The ability of form to be active is the most characteristic feature of mind, and we have something that is mindlike already with the electron."[11]

Similarly, he believes that dividing the universe up into living and nonliving things also has no meaning. Animate and inanimate matter are inseparably interwoven, and life, too, is enfolded throughout the totality of the universe. Even a rock is in some way alive, says Bohm, for life and intelligence are present not only in all of matter, but in "energy," "space," "time," "the fabric of the entire universe," and everything else we abstract out of the holomovement and mistakenly view as separate things.

The idea that consciousness and life (and indeed all things) are ensembles enfolded throughout the universe has an equally dazzling flip side. Just as every portion of a hologram contains the image of the whole, every portion of the universe enfolds the whole. This means that if we knew how to access it we could find the Andromeda galaxy in the thumbnail of our left hand. We could also find Cleopatra meeting Caesar for the first time, for in principle the whole past and implications for the whole future are also enfolded in each small region of space and time. Every cell in our body enfolds the entire cosmos. So does every leaf, every raindrop, and every dust mote, which gives new meaning to William Blake's famous poem:

> To see a World in a Grain of Sand
> And a Heaven in a Wild Flower,
> Hold Infinity in the palm of your hand
> And Eternity in an hour.

The Energy of a Trillion Atomic Bombs in Every Cubic Centimeter of Space

If our universe is only a pale shadow of a deeper order, what else lies hidden, enfolded in the warp and weft of our reality? Bohm has a suggestion. According to our current understanding of physics, every region of space is awash with different kinds of fields composed of waves of varying lengths. Each wave always has at least some energy. When physicists calculate the minimum amount of energy a wave can possess, they find that *every cubic centimeter of empty space contains more energy than the total energy of all the matter in the known universe!*

Some physicists refuse to take this calculation seriously and believe it must somehow be in error. Bohm thinks this infinite ocean of energy does exist and tells us at least a little about the vast and hidden nature of the implicate order. He feels most physicists ignore the existence of this enormous ocean of energy because, like fish who are unaware of the water in which they swim, they have been taught to focus primarily on objects embedded in the ocean, on matter.

Bohm's view that space is as real and rich with process as the matter that moves through it reaches full maturity in his ideas about the implicate sea of energy. Matter does not exist independently from the sea, from so-called empty space. It is a part of space. To explain what he means, Bohm offers the following analogy: A crystal cooled to absolute zero will allow a stream of electrons to pass through it without scattering them. If the temperature is raised, various flaws in the crystal will lose their transparency, so to speak, and begin to scatter electrons. From an electron's point of view such flaws would appear as pieces of "matter" floating in a sea of nothingness, but this is not really the case. The nothingness and the pieces of matter do not exist independently from one another. They are both part of the same fabric, the deeper order of the crystal.

Bohm believes the same is true at our own level of existence. Space is not empty. It is *full,* a plenum as opposed to a vacuum, and is the ground for the existence of everything, including ourselves. The universe is not separate from this cosmic sea of energy, it is a ripple on its surface, a comparatively small "pattern of excitation" in the midst of an unimaginably vast ocean. "This excitation pattern is relatively autonomous and gives rise to approximately recurrent, stable and separable projections into a three-dimensional explicate order of manifestation," states Bohm.[12] In other words, despite its apparent materiality and enormous size, the universe does not exist in and of itself, but is the stepchild of something far vaster and more ineffable. More than that, it is not even a major production of this vaster something, but is only a passing shadow, a mere hiccup in the greater scheme of things.

This infinite sea of energy is not all that is enfolded in the implicate order. Because the implicate order is the foundation that has given birth to everything in our universe, at the very least it also contains every subatomic particle that has been or will be; every configuration of matter, energy, life, and consciousness that is possible, from quasars to the brain of Shakespeare, from the double helix, to the forces that control the sizes and shapes of galaxies. And even this is not all it may contain. Bohm concedes that there is no reason to believe the implicate order is the end of things. There may be other undreamed of orders beyond it, infinite stages of further development.

Experimental Support for Bohm's Holographic Universe

A number of tantalizing findings in physics suggest that Bohm may be correct. Even disregarding the implicate sea of energy, space is filled with light and other electromagnetic waves that constantly crisscross and interfere with one another. As we have seen, all particles are also waves. This means that physical objects and everything else we perceive in reality are composed of interference patterns, a fact that has undeniable holographic implications.

Another compelling piece of evidence comes from a recent experimental finding. In the 1970s the technology became available to actually perform the two-particle experiment outlined by Bell, and a number of different researchers attempted the task. Although the findings were promising, none was able to produce conclusive results. Then in 1982 physicists Alain Aspect, Jean Dalibard and Gérard Roger of the Institute of Optics at the University of Paris succeeded. First they produced a series of twin photons by heating calcium atoms with lasers. Then they allowed each photon to travel in opposite directions through 6.5 meters of pipe and pass through special filters that directed them toward one of two possible polarization analyzers. It took each filter 10 billionths of a second to switch between one analyzer or the other, about 30 billionths of a second less than it took for light to travel the entire 13 meters separating each set of photons. In this way Aspect and his colleagues were able to rule out any possibility of the photons communicating through any known physical process.

Aspect and his team discovered that, as quantum theory predicted, each photon was still able to correlate its angle of polarization with that of its twin. This meant that either Einstein's ban against faster-than-light communication was being violated, or the two photons were nonlocally connected. Because most physicists are opposed to admitting faster-than-light processes into physics, Aspect's experiment is generally viewed as virtual proof that the connection between the two photons is nonlocal. Furthermore, as physicist Paul Davis of the University of Newcastle upon Tyne, England, observes, since *all* particles are continually interacting and separating, "the nonlocal aspects of quantum systems is therefore a general property of nature."[13]

Aspect's findings do not prove that Bohm's model of the universe is correct, but they do provide it with tremendous support. Indeed, as mentioned, Bohm does not believe any theory is correct in an absolute sense, including his own. All are only approximations of the truth, finite maps we use to try to chart territory that is both infinite and indivisible. This does not mean he feels his theory is not testable. He is confident that at some point in the future techniques will be developed which will allow his ideas to be tested (when Bohm is criticized on this point he notes that there are a number of theories in physics, such as "superstring theory," which will probably not be testable for several decades).

The Reaction of the Physics Community

Most physicists are skeptical of Bohm's ideas. For example, Yale physicist Lee Smolin simply does not find Bohm's theory "very compelling, physically."[14] Nonetheless, there is an almost universal respect for Bohm's intelligence. The opinion of Boston University physicist Abner Shimony is representative of this view. "I'm afraid I just don't understand his theory. It is certainly a metaphor and the question is how literally to take the metaphor. Still, he has really thought very deeply about the matter and I think he's done a tremendous service by bringing these questions to the forefront of physics's research instead of just having them swept under the rug. He's been a courageous, daring, and imaginative man."[15]

Such skepticism notwithstanding, there are also physicists who are sympathetic to Bohm's ideas, including such big guns as Roger Penrose of Oxford, the creator of the modern theory of the black hole; Bernard d'Espagnat of the University of Paris, one of the world's leading authorities on the conceptual foundations of quantum theory; and Cambridge's Brian Josephson, winner of the 1973 Nobel Prize in physics. Josephson believes Bohm's implicate order may someday even lead to the inclusion of God or Mind within the framework of science, an idea Josephson supports.[16]

Pribram and Bohm Together

Considered together, Bohm and Pribram's theories provide a profound new way of looking at the world: *Our brains mathematically construct objective reality by interpreting frequencies that are ultimately projections from another dimension, a deeper order of existence that is beyond both space and time: The brain is a hologram enfolded in a holographic universe.*

For Pribram, this synthesis made him realize that the objective world does not exist, at least not in the way we are accustomed to believing. What is "out there" is a vast ocean of waves and frequencies, and reality looks concrete to us only because our brains are able to take this holographic blur and convert it into the sticks and stones and other familiar objects that make up our world. How is the brain (which itself is composed of frequencies of matter) able to take something as insubstantial as a blur of frequencies and make it seem solid to the touch? "The kind of mathematical process that Bekesy simulated with his vibrators is basic to how our brains construct our image of a world out there," Pribram states.[17] In other words, the smoothness of a piece of fine china and the feel of beach sand beneath our feet are really just elaborate versions of the phantom limb syndrome.

According to Pribram this does not mean there aren't china cups and grains of beach sand out there. It simply means that a china cup has two very different aspects to its reality. When it is filtered through the lens of our brain it manifests as a cup. But if we could get rid of our lenses, we'd experience it as an interference pattern. Which one is real and which is illusion? "Both are real to me," says Pribram, "or, if you want to say, neither of them are real."[18]

This state of affairs is not limited to china cups. We, too, have two very different aspects to our reality. We can view ourselves as physical bodies moving through space. Or we can view ourselves as a blur of interference patterns enfolded throughout the cosmic hologram. Bohm believes this second point of view might even be the more correct, for to think of ourselves as a holographic mind/brain *looking* at a holographic universe is again an abstraction, an attempt to separate two things that ultimately cannot be separated.[19]

Do not be troubled if this is difficult to grasp. It is relatively easy to understand the idea of holism in something that is external to us, like an apple in a hologram. What makes it difficult is that in this case we are not looking at the hologram. We are part of the hologram.

The difficulty is also another indication of how radical a revision Bohm and Pribram are trying to make in our way of thinking. But it is not the only radical revision. Pribram's assertion that our brains construct objects pales beside another of Bohm's conclusions: *that we even construct space and time.*[20] The implications of this view are just one of the subjects that will be examined as we explore the effect Bohm and Pribram's ideas have had on the work of researchers in other fields.

PART II

MIND AND BODY

If we were to look closely at an individual human being, we would immediately notice that it is a unique hologram unto itself; self-contained, self-generating, and self-knowledgeable. Yet if we were to remove this being from its planetary context, we would quickly realize that the human form is not unlike a mandala or symbolic poem, for within its form and flow lives comprehensive information about various physical, social, psychological, and evolutionary contexts within which it was created.

—Dr. Ken Dychtwald
in *The Holographic Paradigm*
(Ken Wilber, editor)

3

The Holographic Model and Psychology

While the traditional model of psychiatry and psychoanalysis is strictly personalistic and biographical, modern consciousness research has added new levels, realms, and dimensions and shows the human psyche as being essentially commensurate with the whole universe and all of existence.

—Stanislav Grof
Beyond the Brain

One area of research on which the holographic model has had an impact is psychology. This is not surprising, for, as Bohm has pointed out, consciousness itself provides a perfect example of what he means by undivided and flowing movement. The ebb and flow of our consciousness is not precisely definable but can be seen as a deeper and more fundamental reality out of which our thoughts and ideas unfold. In turn, these thoughts and ideas are not unlike the ripples, eddies, and whirlpools that form in a flowing stream, and like the whirlpools in a stream some can recur and persist in a more or less stable way, while others are evanescent and vanish almost as quickly as they appear.

The holographic idea also sheds light on the unexplainable linkages that can sometimes occur between the consciousnesses of two or more individuals. One of the most famous examples of such linkage is embodied in Swiss psychiatrist Carl Jung's concept of a collective unconscious. Early in his career Jung became convinced that the dreams, artwork, fantasies, and hallucinations of his patients often contained symbols and ideas that could not be explained entirely as products of their personal history. Instead, such symbols more closely resembled the images and themes of the world's great mythologies and religions. Jung concluded that myths, dreams, hallucinations, and religious visions all spring from the same source, a collective unconscious that is shared by all people.

One experience that led Jung to this conclusion took place in 1906 and involved the hallucination of a young man suffering from paranoid schizophrenia. One day while making his rounds Jung found the young man standing at a window and staring up at the sun. The man was also moving his head from side to side in a curious manner. When Jung asked him what he was doing he explained that he was looking at the sun's penis, and when he moved his head from side to side, the sun's penis moved and caused the wind to blow.

At the time Jung viewed the man's assertion as the product of a hallucination. But several years later he came across a translation of a two-thousand-year-old Persian religious text that changed his mind. The text consisted of a series of rituals and invocations designed to bring on visions. It described one of the visions and said that if the participant looked at the sun he would see a tube hanging down from it, and when the tube moved from side to side it would cause the wind to blow. Since circumstances made it extremely unlikely that the man had had contact with the text containing the ritual, Jung concluded that the man's vision was not simply a product of his unconscious mind, but had bubbled up from a deeper level, from the collective unconscious of the human race itself. Jung called such images *archetypes* and believed they were so ancient it's as if each of us has the memory of a two-million-year-old man lurking somewhere in the depths of our unconscious minds.

Although Jung's concept of a collective unconscious has had an enormous impact on psychology and is now embraced by untold thousands of psychologists and psychiatrists, our current understanding of the universe provides no mechanism for explaining its existence. The interconnectedness of all things predicted by the holographic model, however, does offer an explanation. In a universe in which all things are infinitely interconnected, all consciousnesses are also interconnected. Despite appearances, we are beings without borders. Or as

Bohm puts it, "Deep down the consciousness of mankind is one."[1]

If each of us has access to the unconscious knowledge of the entire human race, why aren't we all walking encyclopedias? Psychologist Robert M. Anderson, Jr., of the Rensselaer Polytechnic Institute in Troy, New York, believes it is because we are only able to tap into information in the implicate order that is directly relevant to our memories. Anderson calls this selective process *personal resonance* and likens it to the fact that a vibrating tuning fork will resonate with (or set up a vibration in) another tuning fork *only* if the second tuning fork possesses a similar structure, shape, and size. "Due to personal resonance, relatively few of the almost infinite variety of 'images' in the implicate holographic structure of the universe are available to an individual's personal consciousness," says Anderson. "Thus, when enlightened persons glimpsed this unitive consciousness centuries ago, they did not write out relativity theory because they were not studying physics in a context similar to that in which Einstein studied physics."[2]

Dreams and the Holographic Universe

Another researcher who believes Bohm's implicate order has applications in psychology is psychiatrist Montague Ullman, the founder of the Dream Laboratory at the Maimonides Medical Center in Brooklyn, New York, and a professor emeritus of clinical psychiatry at the Albert Einstein College of Medicine, also in New York. Ullman's initial interest in the holographic concept stemmed also from its suggestion that all people are interconnected in the holographic order. He has good reason for his interest. Throughout the 1960s and 1970s he was responsible for many of the ESP dream experiments mentioned in the introduction. Even today the ESP dream studies conducted at Maimonides stand as some of the best empirical evidence that, in our dreams at least, we are able to communicate with one another in ways that cannot presently be explained.

In a typical experiment a paid volunteer who claimed to possess no psychic ability was asked to sleep in a room in the lab while a person in another room concentrated on a randomly selected painting and tried to get the volunteer to dream of the image it contained. Sometimes the results were inconclusive. But other times the volunteers had dreams that were clearly influenced by the paintings. For example, when the target painting was Tamayo's *Animals*, a picture depicting two dogs flashing their teeth and howling over a pile of bones, the test subject dreamed she was at a banquet where there was not enough meat and everyone was warily eyeing one another as they greedily ate their allotted portions.

In another experiment the target picture was Chagall's *Paris from a Window*, a brightly colored painting depicting a man looking out a window at the Paris skyline. The painting also contained several other unusual features, including a cat with a human face, several small figures of men flying through the air, and a chair covered with flowers. Over the course of several nights the test subject dreamed repeatedly about things French, French architecture, a French policeman's hat, and a man in French attire gazing at various "layers" of a French village. Some of the images in these dreams also appeared to be specific references to the painting's vibrant colors and unusual features, such as the image of a group of bees flying around flowers, and a brightly colored Mardi Gras–type celebration in which the people were wearing costumes and masks.[3]

Although Ullman believes such findings are evidence of the underlying state of interconnectedness Bohm is talking about, he feels that an even more profound example of holographic wholeness can be found in another aspect of dreaming. That is the ability of our dreaming selves often to be far wiser than we ourselves are in our waking state. For instance, Ullman says that in his psychoanalytic practice he could have a patient who seemed completely unenlightened when he was awake—mean, selfish, arrogant, exploitative, and manipulative; a person who had fragmented and dehumanized all of his interpersonal relationships. But no matter how spiritually blind a person may be, or unwilling to recognize his or her own shortcomings, dreams invariably depict their failings honestly and contain metaphors that seem designed to prod him or her gently into a state of greater self-awareness.

Moreover, such dreams were not one-time occurrences. During the course of his practice Ullman noticed that when one of his patients failed to recognize or accept some truth about himself, that truth would surface again and again in his dreams, in different metaphorical guises and linked with different related experiences from his past, but always in an apparent attempt to offer him new opportunities to come to terms with the truth.

Because a man can ignore the counsel of his dreams and still live to be a hundred, Ullman believes this self-monitoring process is striving for more than just the welfare of the individual. He believes that nature is concerned with the survival of the species. He also agrees with Bohm on the importance of wholeness and feels that dreams are nature's way of trying to counteract our seemingly unending compulsion to fragment the world. "An individual can disconnect from all that's cooperative, meaningful, and loving and still survive, but nations don't have that luxury. Unless we learn how to overcome all the ways we've fragmented the human race, nationally, religiously, economically, or whatever, we are going to continue to find ourselves in a position where we can accidentally destroy the whole picture," says Ullman. "The only way we can do that is to look at how we fragment our existence as individuals. Dreams reflect our individual experience, but I think that's because there's a greater underlying need to preserve the species, to maintain species-connectedness."[4]

What is the source of the unending flow of wisdom that bubbles up in our dreams? Ullman admits that he doesn't know, but he offers a suggestion. Given that the implicate order represents in a sense an infinite information source, perhaps it is the origin of this greater fund of knowledge. Perhaps dreams are a bridge between the perceptual and nonmanifest orders and represent a "natural transformation of the implicate into the explicate."[5] If Ullman is correct in this supposition it stands the traditional psychoanalytic view of dreams on its ear, for instead of dream content being something that ascends into consciousness from a primitive substratum of the personality, quite the opposite would be true.

Psychosis and the Implicate Order

Ullman believes that some aspects of psychosis can also be explained by the holographic idea. Both Bohm and Pribram have noted that the experiences mystics have reported throughout the ages—such as feelings of cosmic oneness with the universe, a sense of unity with all life, and so forth—sound very much like descriptions of the implicate order. They suggest that perhaps mystics are somehow able to peer beyond ordinary explicate reality and glimpse its deeper, more holographic qualities. Ullman believes that psychotics are also able to experience certain aspects of the holographic level of reality. But because they are unable to order their experiences rationally, these glimpses are only tragic parodies of the ones reported by mystics.

For example, schizophrenics often report oceanic feelings of one-ness with the universe, but in a magic, delusional way. They describe feeling a loss of boundaries between themselves and others, a belief that leads them to think their thoughts are no longer private. They believe they are able to read the thoughts of others. And instead of viewing people, objects, and concepts as individual things, they often view them as members of larger and larger subclasses, a tendency that seems to be a way of expressing the holographic quality of the reality in which they find themselves.

Ullman believes that schizophrenics try to convey their sense of unbroken wholeness in the way they view space and time. Studies have shown that schizophrenics often treat the converse of any relation as identical to the relation.[6] For instance, according to the schizophrenic's way of thinking, saying that "event A follows event B" is the same as saying "event B follows event A." The idea of one event following another in any kind of time sequence is meaningless, for all points in time are viewed equal. The same is true of spatial relations. If a man's head is above his shoulders, then his shoulders are also above his head. Like the image in a piece of holographic film, things no longer have precise locations, and spatial relationships cease to have meaning.

Ullman believes that certain aspects of holographic thinking are even more pronounced in manic-depressives. Whereas the schizo-phrenic only gets whiffs of the holographic order, the manic is deeply involved in it and grandiosely identifies with its infinite potential. "He can't keep up with all the thoughts and ideas that come at him in so overwhelming a way," states Ullman. "He has to lie, dissemble, and manipulate those about him so as to accommodate to his expansive vista. The end result, of course, is mostly chaos and confusion mixed with occasional outbursts of creativity and success in consensual real-ity."[7] In turn, the manic becomes depressed after he returns from this surreal vacation and once again faces the hazards and chance occur-rences of everyday life.

If it is true that we all encounter aspects of the implicate order when we dream, why don't these encounters have the same effect on us as they do on psychotics? One reason, says Ullman, is that we leave the unique and challenging logic of the dream behind when we wake. Because of his condition the psychotic is forced to contend with it while simultaneously trying to function in everyday reality. Ullman also theorizes that when we dream, most of us have a natural protective mechanism that keeps us from coming into contact with more of the implicate order than we can cope with.

Lucid Dreams and Parallel Universes

In recent years psychologists have become increasingly interested in *lucid dreams*, a type of dream in which the dreamer maintains full waking consciousness and is aware that he or she is dreaming. In addition to the consciousness factor, lucid dreams are unique in several other ways. Unlike normal dreams in which the dreamer is primarily a passive participant, in a lucid dream the dreamer is often able to control the dream in various ways—turn nightmares into pleasant experiences, change the setting of the dream, and/or summon up particular individuals or situations. Lucid dreams are also much more vivid and suffused with vitality than normal dreams. In a lucid dream marble floors seem eerily solid and real, flowers, dazzlingly colorful and fragrant, and everything is vibrant and strangely energized. Researchers studying lucid dreams believe they may lead to new ways to stimulate personal growth, enhance self-confidence, promote mental and physical health, and facilitate creative problem solving.[8]

At the 1987 annual meeting of the Association for the Study of Dreams held in Washington, D.C., physicist Fred Alan Wolf delivered a talk in which he asserted that the holographic model may help explain this unusual phenomenon. Wolf, an occasional lucid dreamer himself, points out that a piece of holographic film actually generates two images, a virtual image that appears to be in the space behind the film, and a real image that comes into focus in the space in front of the film. One difference between the two is that the light waves that compose a virtual image seem to be diverging *from* an apparent focus or source. As we have seen, this is an illusion, for the virtual image of a hologram has no more extension in space than does the image in a mirror. But the real image of a hologram is formed by light waves that are coming *to* a focus, and this is not an illusion. The real image does possess extension in space. Unfortunately, little attention is paid to this real image in the usual applications of holography because an image that comes into focus in empty air is invisible and can only be seen when dust particles pass through it, or when someone blows a puff of smoke through it.

Wolf believes that all dreams are internal holograms, and ordinary dreams are less vivid because they are virtual images. However, he thinks the brain also has the ability to generate real images, and that is exactly what it does when we are dreaming lucidly. The unusual vibrancy of the lucid dream is due to the fact that the waves are converging and not diverging. "If there is a 'viewer' where these waves focus, that viewer will be bathed in the scene, and the scene coming to a focus will 'contain' him. In this way the dream experience will appear 'lucid,' " observes Wolf.[9]

Like Pribram, Wolf believes our minds create the illusion of reality "out there" through the same kind of processes studied by Bekesy. He believes these processes are also what allows the lucid dreamer to create subjective realities in which things like marble floors and flowers are as tangible and real as their so-called objective counterparts. In fact, he thinks our ability to be lucid in our dreams suggests that there may not be much difference between the world at large and the world inside our heads. "When the observer and the observed can separate and say this is the observed and this is the observer, which is an effect one seems to be having when lucid, then I think it's questionable whether [lucid dreams] should be considered subjective," says Wolf.[10]

Wolf postulates that lucid dreams (and perhaps all dreams) are actually visits to parallel universes. They are just smaller holograms within the larger and more inclusive cosmic hologram. He even suggests that the ability to lucid-dream might better be called parallel universe awareness. "I call it parallel universe awareness because I believe that parallel universes arise as other images in the hologram," Wolf states.[11] This and other similar ideas about the ultimate nature of dreaming will be explored in greater depth later in the book.

Hitching a Ride on the Infinite Subway

The idea that we are able to access images from the collective unconscious, or even visit parallel dream universes, pales beside the conclusions of another prominent researcher who has been influenced by the holographic model. He is Stanislav Grof, chief of psychiatric research at the Maryland Psychiatric Research Center and an assistant professor of psychiatry at the Johns Hopkins University School of Medicine.

After more than thirty years of studying nonordinary states of consciousness, Grof has concluded that the avenues of exploration available to our psyches via holographic interconnectedness are more than vast. They are virtually endless.

Grof first became interested in nonordinary states of consciousness in the 1950s while investigating the clinical uses of the hallucinogen LSD at the Psychiatric Research Institute in his native Prague, Czechoslovakia. The purpose of his research was to determine whether LSD had any therapeutic applications. When Grof began his research, most scientists viewed the LSD experience as little more than a stress reaction, the brain's way of responding to a noxious chemical. But when Grof studied the records of his patient's experiences he did not find evidence of any recurring stress reaction. Instead, there was a definite continuity running through each of the patient's sessions. "Rather than being unrelated and random, the experiential content seemed to represent a successive unfolding of deeper and deeper levels of the unconscious," says Grof.[12] This suggested that repeated LSD sessions had important ramifications for the practice and theory of psychotherapy, and provided Grof and his colleagues with the impetus they needed to continue the research. The results were striking. It quickly became clear that serial LSD sessions were able to expedite the psychotherapeutic process and shorten the time necessary for the treatment of many disorders. Traumatic memories that had haunted individuals for years were unearthed and dealt with, and sometimes even serious conditions, such as schizophrenia, were cured.[13] But what was even more startling was that many of the patients rapidly moved beyond issues involving their illnesses and into areas that were uncharted by Western psychology.

One common experience was the reliving of what it was like to be in the womb. At first Grof thought these were just imagined experiences, but as the evidence continued to amass he realized that the knowledge of embryology inherent in the descriptions was often far superior to the patients' previous education in the area. Patients accurately described certain characteristics of the heart sounds of their mother, the nature of acoustic phenomena in the peritoneal cavity, specific details concerning blood circulation in the placenta, and even details about the various cellular and biochemical processes taking place. They also described important thoughts and feelings their mother had had during pregnancy and events such as physical traumas she had experienced.

Whenever possible Grof investigated these assertions, and on several occasions was able to verify them by questioning the mother and other individuals involved. Psychiatrists, psychologists, and biologists who experienced prebirth memories during their training for the program (all the therapists who participated in the study also had to undergo several sessions of LSD psychotherapy) expressed similar astonishment at the apparent authenticity of the experiences.[14]

Most disconcerting of all were those experiences in which the patient's consciousness appeared to expand beyond the usual boundaries of the ego and explore what it was like to be other living things and even other objects. For example, Grof had one female patient who suddenly became convinced she had assumed the identity of a female prehistoric reptile. She not only gave a richly detailed description of what it felt like to be encapsuled in such a form, but noted that the portion of the male of the species' anatomy she found most sexually arousing was a patch of colored scales on the side of its head. Although the woman had no prior knowledge of such things, a conversation Grof had with a zoologist later confirmed that in certain species of reptiles, colored areas on the head do indeed play an important role as triggers of sexual arousal.

Patients were also able to tap into the consciousness of their relatives and ancestors. One woman experienced what it was like to be her mother at the age of three and accurately described a frightening event that had befallen her mother at the time. The woman also gave a precise description of the house her mother had lived in as well as the white pinafore she had been wearing—all details her mother later confirmed and admitted she had never talked about before. Other patients gave equally accurate descriptions of events that had befallen ancestors who had lived decades and even centuries before.

Other experiences included the accessing of racial and collective memories. Individuals of Slavic origin experienced what it was like to participate in the conquests of Genghis Khan's Mongolian hordes, to dance in trance with the Kalahari bushmen, to undergo the initiation rites of the Australian aborigines, and to die as sacrificial victims of the Aztecs. And again the descriptions frequently contained obscure historical facts and a degree of knowledge that was often completely at odds with the patient's education, race, and previous exposure to the subject. For instance, one uneducated patient gave a richly detailed account of the techniques involved in the Egyptian practice of embalming and mummification, including the form and meaning of various amulets and sepulchral boxes, a list of the materials used in the fixing of the mummy cloth, the size and shape of the mummy bandages, and other esoteric facets of Egyptian funeral services. Other individuals tuned into the cultures of the Far East and not only gave impressive descriptions of what it was like to have a Japanese, Chinese, or Tibetan psyche, but also related various Taoist or Buddhist teachings.

In fact, there did not seem to be any limit to what Grof's LSD subjects could tap into. They seemed capable of knowing what it was like to be every animal, and even plant, on the tree of evolution. They could experience what it was like to be a blood cell, an atom, a thermonuclear process inside the sun, the consciousness of the entire planet, and even the consciousness of the entire cosmos. More than that, they displayed the ability to transcend space and time, and occasionally they related uncannily accurate precognitive information. In an even stranger vein they sometimes encountered nonhuman intelligences during their cerebral travels, discarnate beings, spirit guides from "higher planes of consciousness," and other suprahuman entities.

On occasion subjects also traveled to what appeared to be other universes and other levels of reality. In one particularly unnerving session a young man suffering from depression found himself in what seemed to be another dimension. It had an eerie luminescence, and although he could not see anyone he sensed that it was crowded with discarnate beings. Suddenly he sensed a presence very close to him, and to his surprise it began to communicate with him telepathically. It asked him to please contact a couple who lived in the Moravian city of Kromeriz and let them know that their son Ladislav was well taken care of and doing all right. It then gave him the couple's name, street address, and telephone number.

The information meant nothing to either Grof or the young man and seemed totally unrelated to the young man's problems and treatment. Still, Grof could not put it out of his mind. "After some hesitation and with mixed feelings, I finally decided to do what certainly would have made me the target of my colleagues' jokes, had they found out," says Grof. "I went to the telephone, dialed the number in Kromeriz, and asked if I could speak with Ladislav. To my astonishment, the woman on the other side of the line started to cry. When she calmed down, she told me with a broken voice: 'Our son is not with us any more; he passed away, we lost him three weeks ago.' "[15]

In the 1960s Grof was offered a position at the Maryland Psychiatric

Research Center and moved to the United States. The center was also doing controlled studies of the psychotherapeutic applications of LSD, and this allowed Grof to continue his research. In addition to examining the effects of repeated LSD sessions on individuals with various mental disorders, the center also studied its effects on "normal" volunteers—doctors, nurses, painters, musicians, philosophers, scientists, priests, and theologians. Again Grof found the same kind of phenomena occurring again and again. It was almost as if LSD provided the human consciousness with access to a kind of infinite subway system, a labyrinth of tunnels and byways that existed in the subterranean reaches of the unconscious, and one that literally connected everything in the universe with everything else.

After personally guiding over three thousand LSD sessions (each lasting at least five hours) and studying the records of more than two thousand sessions conducted by colleagues, Grof became unalterably convinced that something extraordinary was going on. "After years of conceptual struggle and confusion, I have concluded that the data from LSD research indicate an urgent need for a drastic revision of the existing paradigms for psychology, psychiatry, medicine, and possibly science in general," he states. "There is at present little doubt in my mind that our current understanding of the universe, of the nature of reality, and particularly of human beings, is superficial, incorrect, and incomplete."[16]

Grof coined the term *transpersonal* to describe such phenomena, experiences in which the consciousness transcends the customary boundaries of the personality, and in the late 1960s he joined with several other like-minded professionals, including the psychologist and educator Abraham Maslow, to found a new branch of psychology called *transpersonal psychology*.

If our current way of looking at reality cannot account for transpersonal events, what new understanding might take its place? Grof believes it is the holographic model. As he points out, the essential characteristics of transpersonal experiences—the feeling that all boundaries are illusory, the lack of distinction between part and whole, and the interconnectedness of all things—are all qualities one would expect to find in a holographic universe. In addition, he feels the enfolded nature of space and time in the holographic domain explains why transpersonal experiences are not bound by the usual spatial or temporal limitations.

Grof thinks that the almost endless capacity holograms have for information storage and retrieval also accounts for the fact that visions, fantasies, and other "psychological gestalts," all contain an enormous amount of information about an individual's personality. A single image experienced during an LSD session might contain information about a person's attitude toward life in general, a trauma he experienced during childhood, how much self-esteem he has, how he feels about his parents, and how he feels about his marriage—all embodied in the overall metaphor of the scene. Such experiences are holographic in another way, in that each small part of the scene can also contain an entire constellation of information. Thus, free association and other analytical techniques performed on the scene's miniscule details can call forth an additional flood of data about the individual involved.

The composite nature of archetypal images can be modeled by the holographic idea. As Grof observes, holography makes it possible to build up a sequence of exposures, such as pictures of every member of a large family, on the same piece of film. When this is done the developed piece of film will contain the image of an individual that represents not one member of the family, but all of them at the same time. "These genuinely composite images represent an exquisite model of a certain type of transpersonal experience, such as the archetypal images of the Cosmic Man, Woman, Mother, Father, Lover, Trickster, Fool, or Martyr," says Grof.[17]

If each exposure is taken at a slightly different angle, instead of resulting in a composite picture, the piece of film can be used to create a series of holographic images that appear to flow into one another. Grof believes this illustrates another aspect of the visionary experience, namely, the tendency of countless images to unfold in rapid sequence, each one appearing and then dissolving into the next as if by magic. He thinks holography's success at modeling so many different aspects of the archetypal experience suggests that there is a deep link between holographic processes and the way archetypes are produced.

Indeed, Grof feels that evidence of a hidden, holographic order surfaces virtually every time one experiences a nonordinary state of consciousness:

> Bohm's concept of the unfolded and enfolded orders and the idea that certain important aspects of reality are not accessible to experience and study under ordinary circumstances are of direct relevance for the un-

derstanding of unusual states of consciousness. Individuals who have experienced various nonordinary states of consciousness, including well-educated and sophisticated scientists from various disciplines, frequently report that they entered hidden domains of reality that seemed to be authentic and in some sense implicit in, and supraordinated to, everyday reality.[18]

Holotropic Therapy

Perhaps Grof's most remarkable discovery is that the same phenomena reported by individuals who have taken LSD can also be experienced without resorting to drugs of any kind. To this end, Grof and his wife, Christina, have developed a simple, nondrug technique for inducing these *holotropic,* or nonordinary, states of consciousness. They define a holotropic state of consciousness as one in which it is possible to access the holographic labyrinth that connects all aspects of existence. These include one's biological, psychological, racial, and spiritual history, the past, present, and future of the world, other levels of reality, and all the other experiences already discussed in the context of the LSD experience.

The Grofs call their technique *holotropic therapy* and use only rapid and controlled breathing, evocative music, and massage and body work, to induce altered states of consciousness. To date, thousands of individuals have attended their workshops and report experiences that are every bit as spectacular and emotionally profound as those described by subjects of Grof's previous work on LSD. Grof describes his current work and gives a detailed account of his methods in his book *The Adventure of Self-Discovery.*

Vortices of Thought and Multiple Personalities

A number of researchers have used the holographic model to explain various aspects of the thinking process itself. For example, New York psychiatrist Edgar A. Levenson believes the hologram provides a valuable model for understanding the sudden and transformative changes individuals often experience during psychotherapy. He bases his conclusion on the fact that such changes take place no matter what technique or psychoanalytic approach the therapist uses. Hence, he feels all psychoanalytic approaches are purely ceremonial, and change is due to something else entirely.

Levenson believes that something is resonance. A therapist always knows when therapy is going well, he observes. There is a strong feeling that the pieces of an elusive pattern are all about to come together. The therapist is not saying anything new to the patient, but instead seems to be resonating with something the patient already unconsciously knows: "It is as though a huge, three-dimensional, spatially coded representation of the patient's experience develops in the therapy, running through every aspect of his life, his history and his participation with the therapist. At some point there is a kind of 'overload' and everything falls into place."[19]

Levenson believes these three-dimensional representations of experience are holograms buried deep in the patient's psyche, and a resonance of feeling between the therapist and patient causes them to emerge in a process similar to the way a laser of a certain frequency causes an image made with a laser of the same frequency to emerge from a multiple image hologram. "The holographic model suggests a radically new paradigm which might give us a fresh way of perceiving and connecting clinical phenomena which have always been known to be important, but were relegated to the 'art' of psychotherapy," says Levenson. "It offers a possible theoretical template for change and a practical hope of clarifying psychotherapeutic technique."[20]

Psychiatrist David Shainberg, associate dean of the Postgraduate Psychoanalytic Program at the William Alanson White Institute of Psychiatry in New York, feels Bohm's assertion that thoughts are like vortices in a river should be taken literally and explains why our attitudes and beliefs sometimes become fixed and resistant to change. Studies have shown that vortices are often remarkably stable. The Great Red Spot of Jupiter, a giant vortex of gas over 25,000 miles wide, has remained intact since it was first discovered 300 years ago. Shainberg believes this same tendency toward stability is what causes certain vortices of thought (our ideas and opinions) to become occasionally cemented in our consciousness.

He feels the virtual permanence of some vortices is often detrimental to our growth as human beings. A particularly powerful vortex can dominate our behavior and inhibit our ability to assimilate new ideas and information. It can cause us to become repetitious, create blockages in the creative flow of our consciousness, keep us from seeing the wholeness of ourselves, and make us feel disconnected from our species. Shainberg believes that vortices may even explain things like the nuclear arms race: "Look at the nuclear arms race as a vortex arising out of the greed of human beings who are isolated in their separate selves and do not feel the connection to other human beings. They are also feeling a peculiar emptiness and become greedy for everything they can get to fill themselves. Hence nuclear industries proliferate because they provide large amounts of money and the greed is so extensive that such people do not care what might happen from their actions."[21]

Like Bohm, Shainberg believes our consciousness is constantly unfolding out of the implicate order, and when we allow the same vortices to take form repeatedly he feels we are erecting a barrier between ourselves and the endless positive and novel interactions we could be having with this infinite source of all being. To catch a glimmer of what we are missing, he suggests we look at a child. Children have not yet had the time to form vortices, and this is reflected in the open and flexible way they interact with the world. According to Shainberg the sparkling aliveness of a child expresses the very essence of the unfolding-enfolding nature of consciousness when it is unimpeded.

If you want to become aware of your own frozen vortices of thought, Shainberg recommends you pay close attention to the way you behave in conversation. When people with set beliefs converse with others, they try to justify their identities by espousing and defending their opinions. Their judgments seldom change as a result of any new information they encounter, and they show little interest in allowing any real conversational interaction to take place. A person who is open to the flowing nature of consciousness is more willing to see the frozen condition of the relationships imposed by such vortices of thought. They are committed to exploring conversational interactions, rather than endlessly repeating a static litany of opinions. "Human response and articulation of that response, feedback of reactions to that response and the clarifying of the relationships between different responses, are the way human beings participate in the flow of the implicate order," says Shainberg.[22]

Another psychological phenomena that bears several earmarks of the implicate is multiple personality disorder, or MPD. MPD is a bizarre syndrome in which two or more distinct personalities inhabit a single body. Victims of the disorder, or "multiples," often have no awareness of their condition. They do not realize that control of their body is being passed back and forth between different personalities and instead feel they are suffering from some kind of amnesia, confusion, or black-out spells. Most multiples average between eight to thirteen personalities, although so-called super-multiples may have more than a hundred subpersonalities.

One of the most telling statistics regarding multiples is that 97 percent of them have had a history of severe childhood trauma, often in the form of monstrous psychological, physical, and sexual abuse. This has led many researchers to conclude that becoming a multiple is the psyche's way of coping with extraordinary and soul-crushing pain. By dividing up into one or more personalities the psyche is able to parcel out the pain, in a way, and have several personalities bear what would be too much for just one personality to withstand.

In this sense becoming a multiple may be the ultimate example of what Bohm means by fragmentation. It is interesting to note that when the psyche fragments itself, it does not become a collection of broken and jagged-edged shards, but a collection of smaller wholes, complete and self-sustaining with their own traits, motives, and desires. Although these wholes are not identical copies of the original personality, they are related to the dynamics of the original personality, and this in itself suggests that some kind of holographic process is involved.

Bohm's assertion that fragmentation always eventually proves destructive is also apparent in the syndrome. Although becoming a multiple allows a person to survive an otherwise unendurable childhood, it brings with it a host of unpleasant side effects. These may include depression, anxiety and panic attacks, phobias, heart and respiratory problems, unexplained nausea, migrainelike headaches, tendencies toward self-mutilation, and many other mental and physical disorders. Startlingly, but regular as clockwork, most multiples are diagnosed when they are between the ages of twenty-eight and thirty-five, a "coincidence" that suggests that some inner alarm system may be going off at that age, warning them that it is imperative they are diagnosed and thus obtain the help they need. This idea seems borne out by the fact that multiples who reach their forties before they are diagnosed frequently report having the sense that if they did not seek help soon, any chance of recovery would be lost.[23] Despite the temporary advantages the tortured psyche gains by fragmenting itself, it is clear that mental and physical well-being, and perhaps even survival, still depend on wholeness.

Another unusual feature of MPD is that each of a multiple's personalities possesses a different brain-wave pattern. This is surprising, for as Frank Putnam, a National Institutes of Health psychiatrist who has studied this phenomenon, points out, normally a person's brain-wave pattern does not change even in states of extreme emotion. Brain-wave patterns are not the only thing that varies from personality to personality. Blood flow patterns, muscle tone, heart rate, posture, and even allergies can all change as a multiple shifts from one self to the next.

Since brain-wave patterns are not confined to any single neuron or group of neurons, but are a global property of the brain, this too suggests that some kind of holographic process may be at work. Just as a multiple-image hologram can store and project dozens of whole scenes, perhaps the brain hologram can store and call forth a similar multitude of whole personalities. In other words, perhaps what we call "self" is also a hologram, and when the brain of a multiple clicks from one holographic self to the next, these slide-projectorlike shuttlings are reflected in the global changes that take place in brain-wave activity as well as in the body in general (see fig. 10). The physiological changes that occur as a multiple shifts from one personality to the next also have profound implications for the relationship between mind and health, and will be discussed at greater length in the next chapter.

A Flaw in the Fabric of Reality

Another of Jung's great contributions was defining the concept of synchronicity. As mentioned in the introduction, synchronicities are coincidences that are so unusual and so meaningful they could hardly be attributed to chance alone. Each of us has experienced a synchronicity at some point in our lives, such as when we learn a strange new word and then hear it used in a news broadcast a few hours later, or when we think about an obscure subject and then notice other people talking about it.

A few years back I experienced a series of synchronicities involving

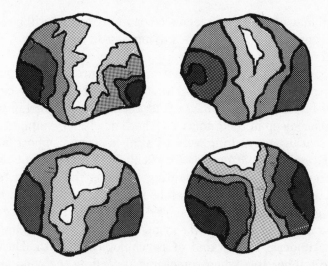

FIGURE 10. The brain-wave patterns of four subpersonalities in an individual suffering from multiple personality disorder. Is it possible that the brain uses holographic principles to store the vast amount of information necessary to house dozens and even hundreds of personalities in a single body? (Redrawn by the author from original art in an article by Bennett G. Braun in the *American Journal of Clinical Hypnosis*)

the rodeo showman Buffalo Bill. Occasionally, while doing a modest workout in the morning before I start writing, I turn on the television. One morning in January 1983, I was doing push-ups while a game show was on, and I suddenly found myself shouting out the name "Buffalo Bill!" At first I was puzzled by my outburst, but then I realized the game-show host had asked the question "What other name was William Frederick Cody known by?" Although I had not been paying conscious attention to the show, for some reason my unconscious mind had zeroed in on this question and had answered it. At the time I did not think much of the occurrence and went about my day. A few hours later a friend telephoned and asked me if I could settle a friendly argument he was having concerning a piece of theater trivia. I offered to try, whereupon my friend asked, "Is it true that John Barrymore's dying words were, 'Aren't you the illegitimate son of Buffalo Bill?' " I thought this second encounter with Buffalo Bill was odd but still chalked it up to coincidence until later that day when a *Smithsonian* magazine arrived in the mail, and I opened it. One of the lead articles was titled "The Last of the Great Scouts Is Back Again." It was about . . . you guessed it: Buffalo Bill. (Incidentally, I was unable to answer my friend's trivia question and still have no idea whether they were Barrymore's dying words or not.)

As incredible as this experience was, the only thing that seemed meaningful about it was its improbable nature. There is, however, another kind of synchronicity that is noteworthy not only because of its improbability, but because of its apparent relationship to events taking place deep in the human psyche. The classic example of this is Jung's scarab story. Jung was treating a woman whose staunchly rational approach to life made it difficult for her to benefit from therapy. After a number of frustrating sessions the woman told Jung about a dream involving a scarab beetle. Jung knew that in Egyptian mythology the scarab represented rebirth and wondered if the woman's unconscious mind was symbolically announcing that she was about to undergo some kind of psychological rebirth. He was just about to tell her this when something tapped on the window, and he looked up to see a gold-green scarab on the other side of the glass (it was the only time a scarab beetle had ever appeared at Jung's window). He opened the window and allowed the scarab to fly into the room as he presented his interpretation of the dream. The woman was so stunned that she tempered her excessive rationality, and from that point on her response to therapy improved.

Jung encountered many such meaningful coincidences during his psychotherapeutic work and noticed that they almost always accompanied periods of emotional intensity and transformation: fundamental changes in belief, sudden and new insights, deaths, births, even changes in profession. He also noticed that they tended to peak when the new realization or insight was just about to surface in a patient's consciousness. As his ideas became more widely known, other therapists began reporting their own experiences with synchronicity.

For example, Zurich-based psychiatrist Carl Alfred Meier, a long-time associate of Jung's, tells of a synchronicity that spanned many years. An American woman suffering from serious depression traveled all the way from Wuchang, China, to be treated by Meier. She was a surgeon and had headed a mission hospital in Wuchang for twenty years. She had also become involved in the culture and was an expert in Chinese philosophy. During the course of her therapy she told Meier of a dream in which she had seen the hospital with one of its wings destroyed. Because her identity was so intertwined with the hospital, Meier felt her dream was telling her she was losing her sense of self, her American identity, and that was the cause of her depression. He advised her to return to the States, and when she did her depression quickly vanished, just as he had predicted. Before she departed he also had her do a detailed sketch of the crumbling hospital.

Years later the Japanese attacked China and bombed Wuchang Hospital. The woman sent Meier a copy of *Life* magazine containing a double-page photograph of the partially destroyed hospital, and it was identical to the drawing she had produced nine years earlier. The symbolic and highly personal message of her dream had somehow spilled beyond the boundaries of her psyche and into physical reality.[24]

Because of their striking nature, Jung became convinced that such synchronicities were not chance occurrences, but were in fact related to the psychological processes of the individuals who experienced them. Since he could not conceive how an occurrence deep in the psyche could *cause* an event or series of events in the physical world, at least in the classical sense, he proposed that some new principle must be involved, an *acausal* connecting principle hitherto unknown to science.

When Jung first advanced this idea, most physicists did not take it seriously (although one eminent physicist of the time, Wolfgang Pauli, felt it was important enough to coauthor a book with Jung on the subject entitled *The Interpretation and Nature of the Psyche*). But now that the existence of nonlocal connections has been established, some physicists are giving Jung's idea another look.* Physicist Paul Davies states, "These *non-local* quantum effects are indeed a form of synchronicity in the sense that they establish a connection—more precisely a correlation—between events for which any form of causal linkage is forbidden."[25]

Another physicist who takes synchronicity seriously is F. David Peat. Peat believes that Jungian-type synchronicities are not only real, but offer further evidence of the implicate order. As we have seen, according to Bohm the apparent separateness of consciousness and matter is an illusion, an artifact that occurs only after both have unfolded into the explicate world of objects and sequential time. If there is no division between mind and matter in the implicate, the ground from which all things spring, then it is not unusual to expect that reality might still be shot through with traces of this deep connectivity. Peat believes that synchronicities are therefore "flaws" in the

---

*As has been mentioned, nonlocal effects are not due to a cause-and-effect relationship and are therefore acausal.

fabric of reality, momentary fissures that allow us a brief glimpse of the immense and unitary order underlying all of nature.

Put another way, Peat thinks that synchronicities reveal the absence of division between the physical world and our inner psychological reality. Thus the relative scarcity of synchronous experiences in our lives shows not only the extent to which we have fragmented ourselves from the general field of consciousness, but also the degree to which we have sealed ourselves off from the infinite and dazzling potential of the deeper orders of mind and reality. According to Peat, when we experience a synchronicity, what we are really experiencing "is the human mind operating, for a moment, in its true order and extending throughout society and nature, moving through orders of increasing subtlety, reaching past the source of mind and matter into creativity itself."[26]

This is an astounding notion. Virtually all of our commonsense prejudices about the world are based on the premise that subjective and objective reality are very much separate. That is why synchronicities seem so baffling and inexplicable to us. But if there is ultimately no division between the physical world and our inner psychological processes, then we must be prepared to change more than just our commonsense understanding of the universe, for the implications are staggering.

One implication is that objective reality is more like a dream than we have previously suspected. For example, imagine dreaming that you are sitting at a table and having an evening meal with your boss and his wife. As you know from experience, all the various props in the dream—the table, the chairs, the plates, and salt and pepper shakers—appear to be separate objects. Imagine also that you experience a synchronicity in the dream; perhaps you are served a particularly unpleasant dish, and when you ask the waiter what it is, he tells you that the name of the dish is Your Boss. Realizing that the unpleasantness of the dish betrays your true feelings about your boss, you become embarrassed and wonder how an aspect of your "inner" self has managed to spill over into the "outer" reality of the scene you are dreaming. Of course, as soon as you wake up you realize the synchronicity was not so strange at all, for there was really no division between your "inner" self and the "outer" reality of the dream. Similarly, you realize that the apparent separateness of the various objects in the dream was also an illusion, for everything was produced by a deeper and more fundamental order—the unbroken wholeness of your own unconscious mind.

If there is no division between the mental and physical worlds, these same qualities are also true of objective reality. According to Peat, this does not mean the material universe is an illusion, because both the implicate and the explicate play a role in creating reality. Nor does it mean that individuality is lost, any more than the image of a rose is lost once it is recorded in a piece of holographic film. It simply means that we are again like vortices in a river, unique but inseparable from the flow of nature. Or as Peat puts it, "the self lives on but as one aspect of the more subtle movement that involves the order of the whole of consciousness."[27]

And so we have come full circle, from the discovery that consciousness contains the whole of objective reality—the entire history of biological life on the planet, the world's religions and mythologies, and the dynamics of both blood cells and stars—to the discovery that the material universe can also contain within its warp and weft the innermost processes of consciousness. Such is the nature of the deep connectivity that exists between all things in a holographic universe. In the next chapter we will explore how this connectivity, as well as other aspects of the holographic idea, affect our current understanding of health.

4

I Sing the Body Holographic

You will hardly know who I am or what I mean,
But I shall be good health to you nevertheless. . . .

—Walt Whitman, "Song of Myself"

A sixty-one-year-old man we'll call Frank was diagnosed as having an almost always fatal form of throat cancer and told he had less than a 5 percent chance of surviving. His weight had dropped from 130 to 98 pounds. He was extremely weak, could barely swallow his own saliva, and was having trouble breathing. Indeed, his doctors had debated whether to give him radiation therapy at all, because there was a distinct possibility the treatment would only add to his discomfort without significantly increasing his chances for survival. They decided to proceed anyway.

Then, to Frank's great good fortune, Dr. O. Carl Simonton, a radiation oncologist and medical director of the Cancer Counseling and Research Center in Dallas, Texas, was asked to participate in his treatment. Simonton suggested that Frank himself could influence the course of his own disease. Simonton then taught Frank a number of relaxation and mental-imagery techniques he and his colleagues had developed. From that point on, three times a day, Frank pictured the radiation he received as consisting of millions of tiny bullets of energy bombarding his cells. He also visualized his cancer cells as weaker and more confused than his normal cells, and thus unable to repair the damage they suffered. Then he visualized his body's white blood cells, the soldiers of the immune system, coming in, swarming over the dead and dying cancer cells, and carrying them to his liver and kidneys to be flushed out of his body.

The results were dramatic and far exceeded what usually happened in such cases when patients were treated solely with radiation. The radiation treatments worked like magic. Frank experienced almost none of the negative side effects—damage to skin and mucous membranes—that normally accompanied such therapy. He regained his lost weight and his strength, and in a mere two months all signs of his cancer had vanished. Simonton believes Frank's remarkable recovery was due in large part to his daily regimen of visualization exercises.

In a follow-up study, Simonton and his colleagues taught their mental-imagery techniques to 159 patients with cancers considered medically incurable. The expected survival time for such a patient is twelve months. Four years later 63 of the patients were still alive. Of those, 14 showed no evidence of disease, the cancers were regressing in 12, and in 17 the disease was stable. The average survival time of the group as a whole was 24.4 months, over twice as long as the national norm.[1]

Simonton has since conducted a number of similar studies, all with positive results. Despite such promising findings, his work is still considered controversial. For instance, critics argue that the individuals who participate in Simonton's studies are not "average" patients. Many of them have sought Simonton out for the express purpose of learning his techniques, and this shows that they already have an extraordinary fighting spirit. Nonetheless, many researchers find Simonton's results compelling enough to support his work, and Simonton himself has set up the Simonton Cancer Center, a successful research and treatment facility in Pacific Palisades, California, devoted to teaching imagery techniques to patients who are fighting various illnesses. The therapeutic use of imagery has also captured the imagination of the public, and a recent survey revealed that it was the fourth most frequently used alternative treatment for cancer.[2]

How is it that an image formed in the mind can have an effect on something as formidable as an incurable cancer? Not surprisingly the holographic theory of the brain can be used to explain this phenomenon as well. Psychologist Jeanne Achterberg, director of research and rehabilitation science at the University of Texas Health Science Center in Dallas, Texas, and one of the scientists who helped develop the imagery techniques Simonton uses, believes it is the holographic imaging capabilities of the brain that provide the key.

As has been noted, all experiences are ultimately just neurophysiological processes taking place in the brain. According to the holographic model the reason we experience some things, such as emotions, as internal realities and others, such as the songs of birds and the barking of dogs, as external realities is because that is where the brain localizes them when it creates the internal hologram that we experience as reality. However, as we have also seen, the brain cannot always distinguish between what is "out there" and what it believes to be "out there," and that is why amputees sometimes have phantom limb sensations. Put another way, in a brain that operates holographically, the remembered image of a thing can have as much impact on the senses as the thing itself.

It can also have an equally powerful effect on the body's physiology, a state of affairs that has been experienced firsthand by anyone who has ever felt their heart race after imagining hugging a loved one. Or anyone who has ever felt their palms grow sweaty after conjuring up the memory of some unusually frightening experience. At first glance the fact that the body cannot always distinguish between an imagined event and a real one may seem strange, but when one takes the holographic model into account—a model that asserts that *all* experiences, whether real or imagined, are reduced to the same common language of holographically organized wave forms—the situation becomes much less puzzling. Or as Achterberg puts it, "When images are regarded in the holographic manner, their omnipotent influence on physical function logically follows. The image, the behavior, and the physiological concomitants are a unified aspect of the same phenomenon."[3]

Bohm uses his idea of the implicate order, the deeper and nonlocal level of existence from which our entire universe springs, to echo the sentiment: "Every action starts from an intention in the implicate order. The imagination is already the creation of the form; it already has the intention and the germs of all the movements needed to carry it out. And it affects the body and so on, so that as creation takes place in that way from the subtler levels of the implicate order, it goes through them until it manifests in the explicate."[4] In other words, in the implicate order, as in the brain itself, imagination and reality are ultimately indistinguishable, and it should therefore come as no surprise to us that images in the mind can ultimately manifest as realities in the physical body.

Achterberg found that the physiological effects produced through the use of imagery are not only powerful, but can also be extremely specific. For example, the term *white blood cell* actually refers to a number of different kinds of cell. In one study, Achterberg decided to see if she could train individuals to increase the number of only one particular type of white blood cell in their body. To do this she taught one group of college students how to image a cell known as a neutrophil, the major constituent of the white blood cell population. She trained a second group to image T-cells, a more specialized kind of white blood cell. At the end of the study the group that learned the neutrophil imagery had a significant increase in the number of neutrophils in their body, but no change in the number of T-cells. The group that learned to image T-cells had a significant increase in the number of that kind of cell, but the number of neutrophils in their body remained the same.[5]

Achterberg says that belief is also critical to a person's health. As she points out, virtually everyone who has had contact with the medical world knows at least one story of a patient who was sent home to die, but because they "believed" otherwise, they astounded their doctors by completely recovering. In her fascinating book *Imagery in Healing* she describes several of her own encounters with such cases. In one, a woman was comatose on admission, paralyzed, and diagnosed with a massive brain tumor. She underwent surgery to "debulk" her tumor (remove as much as is safely possible), but because she was considered close to death, she was sent home without receiving either radiation or chemotherapy.

Instead of promptly dying, the woman became stronger by the day. As her biofeedback therapist, Achterberg was able to monitor the woman's progress, and by the end of sixteen months the woman showed no evidence of cancer. Why? Although the woman was intelligent in a worldly sense, she was only moderately educated and did not really know the meaning of the word *tumor*—or the death sentence it imparted. Hence, she did not believe she was going to die and overcame her cancer with the same confidence and determination she'd used to overcome every other illness in her life, says Achterberg. When Achterberg saw her last, the woman no longer had any traces of paralysis, had thrown away her leg braces and her cane, and had even been out dancing a couple of times.[6]

Achterberg backs up her claim by noting that the mentally retarded and the emotionally disturbed—individuals who cannot comprehend the death sentence society attaches to cancer—also have a significantly lower cancer rate. Over a four-year period in Texas, only about 4 percent of the deaths in these two groups were from cancer, compared to the state norm which was 15 to 18 percent. Intriguingly, there was not one recorded case of leukemia between the years 1925 and 1978 in these two groups. Studies have reported similar results in the United States as a whole, as well as in various other countries including England, Greece, and Romania.[7]

Because of these and other findings Achterberg thinks that a person with an illness, even a common cold, should recruit as many "neural holograms" of health as possible—in the form of beliefs, images of well-being and harmony, and images of specific immune functions being activated. She feels we must also exorcise any beliefs and images that have negative consequences for our health, and realize that our body holograms are more than just pictures. They contain a host of other kinds of information including intellectual understandings and interpretations, prejudices both conscious and unconscious, fears, hopes, worries, and so on.

Achterberg's recommendation that we rid ourselves of negative images is well taken, for there is evidence that imagery can cause illness as well as cure it. In *Love, Medicine, and Miracles,* Bernie Siegel says he often encounters instances where the mental pictures patients use to describe themselves or their lives seem to play a role in the creation of their conditions. Examples include a mastectomy patient who told him she "needed to get something off her chest"; a patient with multiple myeloma in his backbone who said he "was always considered spineless"; and a man with carcinoma of the larynx whose father punished him as a child by constantly squeezing his throat and telling him to "shut up!"

Sometimes the relationship between the image and the illness is so striking it is difficult to understand why it is not apparent to the individual involved, as in the case of a psychotherapist who had emergency surgery to remove several feet of dead intestine and then told Siegel, "I'm glad you're my surgeon. I've been undergoing teaching analysis. I couldn't handle all the shit that was coming up, or digest the crap in my life."[8] Incidents such as these have convinced Siegel that nearly all diseases originate at least to some degree in the mind, but he does not think this makes them psychosomatic or unreal. He prefers to say they are *soma-significant,* a term coined by Bohm to sum up better the relationship, and derived from the Greek word *soma* meaning "body." That all diseases might have their origin in the mind does not disturb Siegel. He sees it rather as a sign of tremendous hope, an indicator that if one has the power to create sickness, one also has the power to create wellness.

The connection between image and illness is so potent, imagery can even be used to predict a patient's prospects for survival. In another landmark experiment, Simonton, his wife, psychologist Stephanie Matthews-Simonton, Achterberg, and psychologist G. Frank Lawlis performed a battery of blood tests on 126 patients with advanced cancer. Then they subjected the patients to an equally extensive array of psychological tests, including exercises in which the patients were asked to draw images of themselves, their cancers, their treatment, and their immune systems. The blood tests offered some information about the patients' condition, but provided no major revelations. However, the results of the psychological tests, particularly the drawings, were encyclopedias of information about the status of the patients' health. Indeed, simply by analyzing patients' drawings, Achterberg later achieved a 95 percent rate of accuracy in predicting who would die within a few months and who would beat their illness and go into remission.[9]

Basketball Games of the Mind

As incredible as the evidence culled by the above-mentioned researchers is, it is just the tip of the iceberg when it comes to the control the holographic mind has over the physical body. And the practical applications of such control are not limited strictly to matters of health. Numerous studies conducted around the world have shown that imagery also has an enormous effect on physical and athletic performance.

In a recent experiment, psychologist Shlomo Breznitz at Hebrew University, Jerusalem, had several groups of Israeli soldiers march forty kilometers (about twenty-five miles), but gave each group different information. He had some groups march thirty kilometers, and then told them they had another ten to go. He told others they were going to march sixty kilometers, but in reality only marched them forty. He allowed some to see distance markers, and provided no clues to others as to how far they had walked. At the end of the study

Breznitz found that the stress hormone levels in the soldiers' blood always reflected their estimates and not the actual distance they had marched.[10] In other words, *their bodies responded not to reality, but to what they were imaging as reality.*

According to Dr. Charles A. Garfield, a former National Aeronautics and Space Administration (NASA) researcher and current president of the Performance Sciences Institute in Berkeley, California, the Soviets have extensively researched the relationship between imagery and physical performance. In one study a phalanx of world-class Soviet athletes was divided into four groups. The first group spent 100 percent of their training time in training. The second spent 75 percent of their time training and 25 percent of their time visualizing the exact movements and accomplishments they wanted to achieve in their sport. The third spent 50 percent of their time training and 50 percent visualizing, and the fourth spent 25 percent training and 75 percent visualizing. Unbelievably, at the 1980 Winter Games in Lake Placid, New York, the fourth group showed the greatest improvement in performance, followed by groups three, two, and one, in that order.[11]

Garfield, who has spent hundreds of hours interviewing athletes and sports researchers around the world, says that the Soviets have incorporated sophisticated imagery techniques into many of their athletic programs and that they believe mental images act as precursors in the process of generating neuromuscular impulses. Garfield believes imagery works because movement is recorded holographically in the brain. In his book *Peak Performance: Mental Training Techniques of the World's Greatest Athletes,* he states, "These images are holographic and function primarily at the subliminal level. The holographic imaging mechanism enables you to quickly solve spatial problems such as assembling a complex machine, choreographing a dance routine, or running visual images of plays through your mind."[12]

Australian psychologist Alan Richardson has obtained similar results with basketball players. He took three groups of basketball players and tested their ability to make free throws. Then he instructed the first group to spend twenty minutes a day practicing free throws. He told the second group not to practice, and had the third group spend twenty minutes a day visualizing that they were shooting perfect baskets. As might be expected, the group that did nothing showed no improvement. The first group improved 24 percent, but through the power of imagery alone, the third group improved an astonishing 23 percent, almost as much as the group that practiced.[13]

The Lack of Division Between Health and Illness

Physician Larry Dossey believes that imagery is not the only tool the holographic mind can use to effect changes in the body. Another is simply the recognition of the unbroken wholeness of all things. As Dossey observes, we have a tendency to view illness as external to us. Disease comes from without and besieges us, upsetting our well-being. But if space and time, and all other things in the universe, are truly inseparable, then we cannot make a distinction between health and disease.

How can we put this knowledge to practical use in our lives? When we stop seeing illness as something separate and instead view it as part of a larger whole, as a milieu of behavior, diet, sleep, exercise patterns, and various other relationships with the world at large, we often get better, says Dossey. As evidence he calls attention to a study in which chronic headache sufferers were asked to keep a diary of the frequency and severity of their headaches. Although the record was intended to be a first step in preparing the headache sufferers for further treatment, most of the subjects found that when they began to keep a diary, their headaches disappeared![14]

In another experiment cited by Dossey, a group of epileptic children and their families were videotaped as they interacted with one another. Occasionally, there were emotional outbursts during the sessions, which were often followed by actual seizures. When the children were shown the tapes and saw the relationship between these emotional events and their seizures, they became almost seizure-free.[15] Why? By keeping a diary or watching a videotape, the subjects were able to see their condition in relationship to the larger pattern of their lives. When this happens, illness can no longer be viewed "as an intruding disease originating elsewhere, but as part of a process of living which can accurately be described as an unbroken whole," says Dossey. "When our focus is toward a principle of relatedness and oneness, and away from fragmentation and isolation, health ensues."[16]

Dossey feels the word *patient* is as misleading as the word *particle*. Instead of being separate and fundamentally isolated biological units, we are essentially dynamic processes and patterns that are no more analyzable into parts than are electrons. More than this, we are connected, connected to the forces that create both sickness and health, to the beliefs of our society, to the attitudes of our friends, our family, and our doctors, and to the images, beliefs, and even the very words we use to apprehend the universe.

In a holographic universe we are also connected to our bodies, and in the preceding pages we have seen some of the ways these connections manifest themselves. But there are others, perhaps even an infinity of others. As Pribram states, "If indeed every part of our body is a reflection of the whole, then there must be all kinds of mechanisms to control what's going on. Nothing is firm at this point."[17] Given our ignorance in the matter, instead of asking *how* the mind controls the body holographic, perhaps a more important question is, What is the extent of this control? Are there any limitations on it, and if so, what are they? That is the question to which we now turn our attention.

The Healing Power of Nothing at All

Another medical phenomenon that provides us with a tantalizing glimpse of the control the mind has over the body is the placebo effect. A placebo is any medical treatment that has no specific action on the body but is given either to humor a patient, or as a control in a double-blind experiment, that is, a study in which one group of individuals is given a real treatment and another group is given a fake treatment. In such experiments neither the researchers nor the individuals being tested know which group they are in so that the effects of the real treatment can be assessed more accurately. Sugar pills are often used as placebos in drug studies. So is saline solution (distilled water with salt in it), although placebos need not always be drugs. Many believe that any medical benefit derived from crystals, copper bracelets, and other nontraditional remedies is also due to the placebo effect.

Even surgery has been used as a placebo. In the 1950s, angina pectoris, recurrent pain in the chest and left arm due to decreased blood flow to the heart, was commonly treated with surgery. Then some resourceful doctors decided to conduct an experiment. Rather than perform the customary surgery, which involved tying off the mammary artery, they cut patients open and then simply sewed them back up again. The patients who received the sham surgery reported just as much relief as the patients who had the full surgery. The full surgery, as it turned out, was only producing a placebo effect.[18] Nonetheless, the success of the sham surgery indicates that somewhere deep in all of us we have the ability to control angina pectoris.

And that is not all. In the last half century the placebo effect has been extensively researched in hundreds of different studies around the world. We now know that on average 35 percent of all people who receive a given placebo will experience a significant effect, although this number can vary greatly from situation to situation. In addition to angina pectoris, conditions that have proved responsive to placebo treatment include migraine headaches, allergies, fever, the common cold, acne, asthma, warts, various kinds of pain, nausea and seasickness, peptic ulcers, psychiatric syndromes such as depression and anxiety, rheumatoid and degenerative arthritis, diabetes, radiation sickness, Parkinsonism, multiple sclerosis, and cancer.

Clearly these range from the not so serious to the life threatening, but placebo effects on even the mildest conditions may involve physiological changes that are near miraculous. Take, for example, the lowly wart. Warts are a small tumorous growth on the skin caused by a virus. They are also extremely easy to cure through the use of placebos, as is evidenced by the nearly endless folk rituals—ritual itself being a kind of placebo—that are used by various cultures to get rid of them. Lewis Thomas, president emeritus of Memorial Sloan-Kettering Cancer Center in New York, tells of one physician who regularly rid his patients of warts simply by painting a harmless purple dye on them. Thomas feels that explaining this small miracle by saying it's just the unconscious mind at work doesn't begin to do the placebo effect justice. "If my unconscious can figure out how to manipulate the mechanisms needed for getting around that virus, and for deploying all the various cells in the correct order for tissue rejection, then all I have to say is that my unconscious is a lot further along than I am," he states.[19]

The effectiveness of a placebo in any given circumstance also varies greatly. In nine double-blind studies comparing placebos to aspirin, placebos proved to be 54 percent as effective as the actual analgesic.[20] From this one might expect that placebos would be even less effective when compared to a much stronger painkiller such as morphine, but this is not the case. In six double-blind studies placebos were found to be 56 percent as effective as morphine in relieving pain![21]

Why? One factor that can affect the effectiveness of a placebo is the method in which it is given. Injections are generally perceived as more

potent than pills, and hence giving a placebo in an injection can enhance its effectiveness. Similarly, capsules are often seen as more
effective than tablets, and even the size, shape, and color of a pill can
play a role. In a study designed to determine the suggestive value of
a pill's color, researchers found that people tend to view yellow or
orange pills as mood manipulators, either stimulants or depressants.
Dark red pills are assumed to be sedatives; lavender pills, hallucino-
gens; and white pills, painkillers.[22]

Another factor is the attitude the doctor conveys when he prescribes
the placebo. Dr. David Sobel, a placebo specialist at Kaiser Hospital,
California, relates the story of a doctor treating an asthma patient
who was having an unusually difficult time keeping his bronchial tubes
open. The doctor ordered a sample of a potent new medicine from a
pharmaceutical company and gave it to the man. Within minutes the
man showed spectacular improvement and breathed more easily. However, the next time he had an attack, the doctor decided to see what
would happen if he gave the man a placebo. This time the man complained that there must be something wrong with the prescription
because it didn't completely eliminate his breathing difficulty. This
convinced the doctor that the sample drug was indeed a potent new
asthma medication—until he received a letter from the pharmaceutical
company informing him that instead of the new drug, they had accidentally sent him a placebo! Apparently it was the doctor's unwitting
enthusiasm for the first placebo, and not the second, that accounted
for the discrepancy.[23]

In terms of the holographic model, the man's remarkable response
to the placebo asthma medication can again be explained by the mind/
body's ultimate inability to distinguish between an imagined reality
and a real one. The man believed he was being given a powerful new
asthma drug, and this belief had as dramatic a physiological effect on
his lungs as if he had been given a real drug. Achterberg's warning
that the neural holograms that impact on our health are varied and
multifaceted is also underscored by the fact that even something as
subtle as the doctor's slightly different attitude (and perhaps body
language) while administering the two placebos was enough to cause
one to work and the other to fail. It is clear from this that even
information received subliminally can contribute greatly to the beliefs
and mental images that impact on our health. One wonders how many
drugs have worked (or not worked) because of the attitude the doctor
conveyed while administering them.

Tumors That Melt Like Snowballs on a Hot Stove

Understanding the role such factors play in a placebo's effectiveness is important, for it shows how our ability to control the body holographic is molded by our beliefs. Our minds have the power to get rid of warts, to clear our bronchial tubes, and to mimic the painkilling ability of morphine, but because we are unaware that we possess the power, we must be fooled into using it. This might almost be comic if it were not for the tragedies that often result from our ignorance of our own power.

No incident better illustrates this than a now famous case reported by psychologist Bruno Klopfer. Klopfer was treating a man named Wright who had advanced cancer of the lymph nodes. All standard treatments had been exhausted, and Wright appeared to have little time left. His neck, armpits, chest, abdomen, and groin were filled with tumors the size of oranges, and his spleen and liver were so enlarged that two quarts of milky fluid had to be drained out of his chest every day.

But Wright did not want to die. He had heard about an exciting new drug called Krebiozen, and he begged his doctor to let him try it. At first his doctor refused because the drug was only being tried on people with a life expectancy of at least three months. But Wright was so unrelenting in his entreaties, his doctor finally gave in. He gave Wright an injection of Krebiozen on Friday, but in his heart of hearts he did not expect Wright to last the weekend. Then the doctor went home.

To his surprise, on the following Monday he found Wright out of bed and walking around. Klopfer reported that his tumors had "melted like snowballs on a hot stove" and were half their original size. This was a far more rapid decrease in size than even the strongest X-ray treatments could have accomplished. Ten days after Wright's first Krebiozen treatment, he left the hospital and was, as far as his doctors could tell, cancer free. When he had entered the hospital he had needed an oxygen mask to breathe, but when he left he was well enough to fly his own plane at 12,000 feet with no discomfort.

Wright remained well for about two months, but then articles began to appear asserting that Krebiozen actually had no effect on cancer of the lymph nodes. Wright, who was rigidly logical and scientific in his thinking, became very depressed, suffered a relapse, and was readmitted to the hospital. This time his physician decided to try an experiment. He told Wright that Krebiozen was every bit as effective as it had seemed, but that some of the initial supplies of the drug had deteriorated during shipping. He explained, however, that he had a new highly concentrated version of the drug and could treat Wright with this. Of course the physician did not have a new version of the drug and intended to inject Wright with plain water. To create the proper atmosphere he even went through an elaborate procedure before injecting Wright with the placebo.

Again the results were dramatic. Tumor masses melted, chest fluid vanished, and Wright was quickly back on his feet and feeling great. He remained symptom-free for another two months, but then the American Medical Association announced that a nationwide study of Krebiozen had found the drug worthless in the treatment of cancer. This time Wright's faith was completely shattered. His cancer blossomed anew and he died two days later.[24]

Wright's story is tragic, but it contains a powerful message: When we are fortunate enough to bypass our disbelief and tap the healing forces within us, we can cause tumors to melt away overnight.

In the case of Krebiozen only one person was involved, but there are similar cases involving many more people. Take a chemotherapeutic agent called cis-platinum. When cis-platinum first became available it, too, was touted as a wonder drug, and 75 percent of the people who received it benefited from the treatment. But after the initial wave of excitement and the use of cis-platinum became more routine, its rate of effectiveness dropped to about 25 to 30 percent. Apparently most of the benefit obtained from cis-platinum was due to the placebo effect.[25]

Do Any Drugs Really Work?

Such incidents raise an important question. If drugs such as Krebiozen and cis-platinum work when we believe in them and stop working when we stop believing in them, what does this imply about the nature of drugs in general? This is a difficult question to answer, but we do have some clues. For instance, physician Herbert Benson of Harvard Medical School points out that the vast majority of treatments prescribed prior to this century, from leeching to consuming lizard's blood, were useless, but because of the placebo effect, they were no doubt helpful at least some of the time.[26]

Benson, along with Dr. David P. McCallie, Jr., of Harvard's Thorn-
dike Laboratory, reviewed studies of various treatments for angina
pectoris that have been prescribed over the years and discovered that
although remedies have come and gone, the success rates—even for
treatments that are now discredited—have always remained high.[27]
From these two observations it is evident that the placebo effect has
played an important role in medicine in the past, but does it still play
a role today? The answer, it seems, is yes. The federal Office of Technology Assessment estimates that more than 75 percent of all current
medical treatments have not been subjected to sufficient scientific
scrutiny, a figure that suggests that doctors may still be giving placebos and not know it (Benson, for one, believes that, at the very least,
many over-the-counter medications act primarily as placebos).[28]

Given the evidence we have looked at so far, one might almost
wonder if all drugs are placebos. Clearly the answer is no. Many drugs
are effective whether we believe in them or not: Vitamin C gets rid of
scurvy, and insulin makes diabetics better even when they are skeptical. But still the issue is not quite as clear-cut as it may seem. Consider
the following.

In a 1962 experiment Drs. Harriet Linton and Robert Langs told test
subjects they were going to participate in a study of the effects of
LSD, but then gave them a placebo instead. Nonetheless, half an hour
after taking the placebo, the subjects began to experience the classic
symptoms of the actual drug, loss of control, supposed insight into the
meaning of existence, and so on. These "placebo trips" lasted several
hours.[29]

A few years later, in 1966, the now infamous Harvard psychologist
Richard Alpert journeyed to the East to look for holy men who could
offer him insight into the LSD experience. He found several who were
willing to sample the drug and, interestingly, received a variety of
reactions. One pundit told him it was good, but not as good as meditation. Another, a Tibetan lama, complained that it only gave him a
headache.

But the reaction that fascinated Alpert most came from a wizened
little holy man in the foothills of the Himalayas. Because the man was
over sixty, Alpert's first inclination was to give him a gentle dose of
50 to 75 micrograms. But the man was much more interested in one
of the 305 microgram pills Alpert had brought with him, a relatively
sizable dose. Reluctantly, Alpert gave him one of the pills, but still the
man was not satisfied. With a twinkle in his eye he requested another

and then another and placed all 915 micrograms of LSD on his tongue, a massive dose by any standard, and swallowed them (in comparison, the average dose Grof used in his studies was about 200 micrograms).

Aghast, Alpert watched intently, expecting the man to start waving his arms and whooping like a banshee, but instead he behaved as if nothing had happened. He remained that way for the rest of the day, his demeanor as serene and unperturbed as it always was, save for the twinkling glances he occasionally tossed Alpert. The LSD apparently had little or no effect on him. Alpert was so moved by the experience he gave up LSD, changed his name to Ram Dass, and converted to mysticism.[30]

And so taking a placebo may well produce the same effect as taking the real drug, and taking the real drug might produce no effect. This topsy-turvy state of affairs has also been demonstrated in experiments involving amphetamines. In one study, ten subjects were placed in each of two rooms. In the first room, nine were given a stimulating amphetamine and the tenth a sleep-producing barbiturate. In the second room the situation was reversed. In both instances, the person singled out behaved exactly as his companions did. In the first room instead of falling asleep the lone barbiturate taker became animated and speedy, and in the second room the lone amphetamine taker fell asleep.[31] There is also a case on record of a man addicted to the stimulant Ritalin, whose addiction is then transferred to a placebo. In other words, the man's doctor enabled him to avoid all the usual unpleasantries of Ritalin withdrawal by secretly replacing his prescription with sugar pills. Unfortunately the man then went on to display an addiction to the placebo![32]

Such events are not limited to experimental situations. Placebos also play a role in our everyday lives. Does caffeine keep you awake at night? Research has shown that even an injection of caffeine won't keep caffeine-sensitive individuals awake if they believe they are receiving a sedative.[33] Has an antibiotic ever helped you get over a cold or sore throat? If so, you were experiencing the placebo effect. All colds are caused by viruses, as are several types of sore throat, and antibiotics are only effective against bacterial infections, not viral infections. Have you ever experienced an unpleasant side effect after taking a medication? In a study of a tranquilizer called mephenesin, researchers found that 10 to 20 percent of the test subjects experienced negative side effects—including nausea, itchy rash, and heart palpitations—regardless of whether they were given the actual drug or a placebo.*[34] Similarly, in a recent study of a new kind of chemo-therapy, 30 percent of the individuals in the *control* group, the group given placebos, lost their hair.[35] So if you know someone who is taking chemotherapy, tell them to try to be optimistic in their expectations. The mind is a powerful thing.

In addition to offering us a glimpse of this power, placebos also support a more holographic approach to understanding the mind/body relationship. As health and nutrition columnist Jane Brody observes in an article in the *New York Times*, "The effectiveness of placebos provides dramatic support for a 'holistic' view of the human organism, a view that is receiving increasing attention in medical research. This view holds that the mind and body continually interact and are too closely interwoven to be treated as independent entities."[36]

The placebo effect may also be affecting us in far vaster ways than we realize, as is evidenced by a recent and extremely puzzling medical mystery. If you have watched any television at all in the last year or so, you have no doubt seen a blitzkrieg of commercials promoting aspirin's ability to decrease the risk of heart attack. There is a good deal of convincing evidence to back this up, otherwise television censors, who are real sticklers for accuracy when it comes to medical claims in commercials, wouldn't allow such copy on the air. This is all well and good. The only problem is that aspirin doesn't seem to have the same effect on people in England. A six-year study of 5,139 British doctors revealed no evidence that aspirin reduces the risk of heart attack.[37] Is there a flaw in somebody's research, or is it possible that some kind of massive placebo effect is to blame? Whatever the case, don't stop believing in the prophylactic benefits of aspirin. It still may save your life.

The Health Implications of Multiple Personality

Another condition that graphically illustrates the mind's power to affect the body is Multiple Personality Disorder (MPD). In addition to possessing different brain-wave patterns, the subpersonalities of a multiple have a strong psychological separation from one another.

*Of course I am by no means suggesting that all drug side effects are the result of the placebo effect. Should you experience a negative reaction to a drug, *always* consult a physician.

Each has his own name, age, memories, and abilities. Often each also has his own style of handwriting, announced gender, cultural and racial background, artistic talents, foreign language fluency, and IQ.

Even more noteworthy are the biological changes that take place in a multiple's body when they switch personalities. Frequently a medical condition possessed by one personality will mysteriously vanish when another personality takes over. Dr. Bennett Braun of the International Society for the Study of Multiple Personality, in Chicago, has documented a case in which all of a patient's subpersonalities were allergic to orange juice, except one. If the man drank orange juice when one of his allergic personalities was in control, he would break out in a terrible rash. But if he switched to his nonallergic personality, the rash would instantly start to fade and he could drink orange juice freely.[38]

Dr. Francine Howland, a Yale psychiatrist who specializes in treating multiples, relates an even more striking incident concerning one multiple's reaction to a wasp sting. On the occasion in question, the man showed up for his scheduled appointment with Howland with his eye completely swollen shut from a wasp sting. Realizing he needed medical attention, Howland called an ophthalmologist. Unfortunately, the soonest the opthalmologist could see the man was an hour later, and because the man was in severe pain, Howland decided to try something. As it turned out, one of the man's alternates was an "anesthetic personality" who felt absolutely no pain. Howland had the anesthetic personality take control of the body, and the pain ended. But something else also happened. By the time the man arrived at his appointment with the ophthalmologist, the swelling was gone and his eye had returned to normal. Seeing no need to treat him, the ophthalmologist sent him home.

After a while, however, the anesthetic personality relinquished control of the body, and the man's original personality returned, along with all the pain and swelling of the wasp sting. The next day he went back to the ophthalmologist to at last be treated. Neither Howland nor her patient had told the ophthalmologist that the man was a multiple, and after treating him, the ophthalmologist telephoned Howland. "He thought time was playing tricks on him." Howland laughed. "He just wanted to make sure that I had actually called him the day before and he had not imagined it."[39]

Allergies are not the only thing multiples can switch on and off. If there was any doubt as to the control the unconscious mind has over drug effects, it is banished by the pharmacological wizardry of the multiple. By changing personalities, a multiple who is drunk can instantly become sober. Different personalities also respond differently to different drugs. Braun records a case in which 5 milligrams of diazepam, a tranquilizer, sedated one personality, while 100 milligrams had little or no effect on another. Often one or several of a multiple's personalities are children, and if an adult personality is given a drug and then a child's personality takes over, the adult dosage may be too much for the child and result in an overdose. It is also difficult to anesthetize some multiples, and there are accounts of multiples waking up on the operating table after one of their "unanesthetizable" subpersonalities has taken over.

Other conditions that can vary from personality to personality include scars, burn marks, cysts, and left- and right-handedness. Visual acuity can differ, and some multiples have to carry two or three different pairs of eyeglasses to accommodate their alternating personalities. One personality can be color-blind and another not, and even eye color can change. There are cases of women who have two or three menstrual periods each month because each of their subpersonalities has its own cycle. Speech pathologist Christy Ludlow has found that the voice pattern for each of a multiple's personalities is different, a feat that requires such a deep physiological change that even the most accomplished actor cannot alter his voice enough to disguise his voice pattern.[40] One multiple, admitted to a hospital for diabetes, baffled her doctors by showing no symptoms when one of her nondiabetic personalities was in control.[41] There are accounts of epilepsy coming and going with changes in personality, and psychologist Robert A. Phillips, Jr., reports that even tumors can appear and disappear (although he does not specify what kind of tumors).[42]

Multiples also tend to heal faster than normal individuals. For example, there are several cases on record of third-degree burns healing with extraordinary rapidity. Most eerie of all, at least one researcher—Dr. Cornelia Wilbur, the therapist whose pioneering treatment of Sybil Dorsett was portrayed in the book *Sybil*—is convinced that multiples don't age as fast as other people.

How could such things be? At a recent symposium on the multiple personality syndrome, a multiple named Cassandra provided a possible answer. Cassandra attributes her own rapid healing ability both to the visualization techniques she practices and to something she calls *parallel processing*. As she explained, even when her alternate personalities are not in control of her body, they are still aware. This enables her to "think" on a multitude of different channels at once, to do things like work on several different term papers simultaneously, and even "sleep" while other personalities prepare her dinner and clean her house.

Hence, whereas normal people only do healing imagery exercises two or three times a day, Cassandra does them around the clock. She even has a subpersonality named Celese who possesses a thorough knowledge of anatomy and physiology, and whose sole function is to spend twenty-four hours a day meditating and imaging the body's well-being. According to Cassandra, it is this full-time attention to her health that gives her an edge over normal people. Other multiples have made similar claims.[43]

We are deeply attached to the inevitability of things. If we have bad vision, we believe we will have bad vision for life, and if we suffer from diabetes, we do not for a moment think our condition might vanish with a change in mood or thought. But the phenomenon of multiple personality challenges this belief and offers further evidence of just how much our psychological states can affect the body's biology. If the psyche of an individual with MPD is a kind of multiple image hologram, it appears that the body is one as well, and can switch from one biological state to another as rapidly as the flutter of a deck of cards.

The systems of control that must be in place to account for such capacities is mind-boggling and makes our ability to will away a wart look pale. Allergic reaction to a wasp sting is a complex and multifaceted process and involves the organized activity of antibodies, the production of histamine, the dilation and rupture of blood vessels, the excessive release of immune substances, and so on. What unknown pathways of influence enable the mind of a multiple to freeze all these processes in their tracks? Or what allows them to suspend the effects of alcohol and other drugs in the blood, or turn diabetes on and off? At the moment we don't know and must console ourselves with one simple fact. Once a multiple has undergone therapy and in some way becomes whole again, he or she can still make these switches at will.[44] This suggests that somewhere in our psyches we *all* have the ability to control these things. And still this is not all we can do.

Pregnancy, Organ Transplants, and Tapping the Genetic Level

As we have seen, simple everyday belief can also have a powerful effect on the body. Of course most of us do not have the mental discipline to completely control our beliefs (which is why doctors must use placebos to fool us into tapping the healing forces within us). To regain that control we must first understand the different types of belief that can affect us, for these too offer their own unique window on the plasticity of the mind/body relationship.

CULTURAL BELIEFS

One type of belief is imposed on us by our society. For example, the people of the Trobriand Islands engage freely in sexual relations before marriage, but premarital pregnancy is strongly frowned upon. They use no form of contraception, and seldom if ever resort to abortion. Yet premarital pregnancy is virtually unknown. This suggests that, because of their cultural beliefs, the unmarried women are unconsciously preventing themselves from getting pregnant.[45] There is evidence that something similar may be going on in our own culture. Almost everyone knows of a couple who have tried unsuccessfully for years to have a child. They finally adopt, and shortly thereafter the woman gets pregnant. Again this suggests that finally having a child enabled the woman and/or her husband to overcome some sort of inhibition that was blocking the effects of her and/or his fertility.

The fears we share with the other members of our culture can also affect us greatly. In the nineteenth century, tuberculosis killed tens of thousands of people, but starting in the 1880s, death rates began to plummet. Why? Previous to that decade no one knew what caused TB, which gave it an aura of terrifying mystery. But in 1882 Dr. Robert Koch made the momentous discovery that TB was caused by a bacterium. Once this knowledge reached the general public, death rates fell from 600 per 100,000 to 200 per 100,000, despite the fact that it would be nearly half a century before an effective drug treatment could be found.[46]

Fear apparently has been an important factor in the success rates of organ transplants as well. In the 1950s kidney transplants were only a tantalizing possibility. Then a doctor in Chicago made what seemed to be a successful transplant. He published his findings, and soon after other successful transplants took place around the world. Then the first transplant failed. In fact, the doctor discovered that the kidney had actually been rejected from the start. But it did not matter. Once transplant recipients believed they could survive, they did, and success rates soared beyond all expectations.[47]

THE BELIEFS WE EMBODY IN OUR ATTITUDES

Another way belief manifests in our lives is through our attitudes. Studies have shown that the attitude an expectant mother has toward her baby, and pregnancy in general, has a direct correlation with the complications she will experience during childbirth, as well as with the medical problems her newborn infant will have after it is born.[48] Indeed, in the past decade an avalanche of studies has poured in demonstrating the effect our attitudes have on a host of medical conditions. People who score high on tests designed to measure hostility and aggression are seven times more likely to die from heart problems than people who receive low scores.[49] Married women have stronger immune systems than separated or divorced women, and *happily* married women have even stronger immune systems.[50] People with AIDS who display a fighting spirit live longer than AIDS-infected individuals who have a passive attitude.[51] People with cancer also live longer if they maintain a fighting spirit.[52] Pessimists get more colds than optimists.[53] Stress lowers the immune response;[54] people who have just lost their spouse have an increased incidence of illness and disease,[55] and on and on.

THE BELIEFS WE EXPRESS THROUGH THE POWER OF OUR WILL

The types of belief we have examined so far can be viewed largely as passive beliefs, beliefs we allow our culture or the normal state of our thoughts to impose upon us. Conscious belief in the form of a steely and unswerving will can also be used to sculpt and control the body holographic. In the 1970s, Jack Schwarz, a Dutch-born author and lecturer, astounded researchers in laboratories across the United States with his ability to willfully control his body's internal biological processes.

In studies conducted at the Menninger Foundation, the University of California's Langley Porter Neuropsychiatric Institute, and others,

Schwarz astonished doctors by sticking mammoth six-inch sailmaker's needles completely through his arms without bleeding, without flinching, and without producing beta brain waves (the type of brain waves normally produced when a person is in pain). Even when the needles were removed, Schwarz still did not bleed, and the puncture holes closed tightly. In addition, Schwarz altered his brain-wave rhythms at will, held burning cigarettes against his flesh without harming himself, and even carried live coals around in his hands. He claims he acquired these abilities when he was in a Nazi concentration camp and had to learn how to control pain in order to withstand the terrible beatings he endured. He believes anyone can learn voluntary control of their body and thus gain responsibility for his or her own health.[56]

Oddly enough, in 1947 another Dutchman demonstrated similar abilities. The man's name was Mirin Dajo, and in public performances at the Corso Theater in Zurich, he left audiences stunned. In plain view Dajo would have an assistant stick a fencing foil completely through his body, clearly piercing vital organs but causing Dajo no harm or pain. Like Schwarz, when the foil was removed, Dajo did not bleed and only a faint red line marked the spot where the foil had entered and exited.

Dajo's performance proved so nerve-racking to his audiences that eventually one spectator suffered a heart attack, and Dajo was legally banned from performing in public. However, a Swiss doctor named Hans Naegeli-Osjord learned of Dajo's alleged abilities and asked him if he would submit to scientific scrutiny. Dajo agreed, and on May 31, 1947, he entered the Zurich cantonal hospital. In addition to Dr. Naegeli-Osjord, Dr. Werner Brunner, the chief of surgery at the hospital, was also present, as were numerous other doctors, students, and journalists. Dajo bared his chest and concentrated, and then, in full view of the assemblage, he had his assistant plunge the foil through his body.

As always, no blood flowed and Dajo remained completely at ease. But he was the only one smiling. The rest of the crowd had turned to stone. By all rights, Dajo's vital organs should have been severely damaged, and his seeming good health was almost too much for the doctors to bear. Filled with disbelief, they asked Dajo if he would submit to an X ray. He agreed and without apparent effort accompanied them up the stairs to the X-ray room, the foil still through his abdomen. The X ray was taken and the result was undeniable. Dajo was indeed impaled. Finally, a full twenty minutes after he had been

pierced, the foil was removed, leaving only two faint scars. Later, Dajo
was tested by scientists in Basel, and even let the doctors themselves
run him through with the foil. Dr. Naegeli-Osjord later related the
entire case to the German physicist Alfred Stelter, and Stelter reports
it in his book *Psi-Healing*.[57]

Such supernormal feats of control are not limited to the Dutch. In
the 1960s Gilbert Grosvenor, the president of the National Geographic
Society, his wife, Donna, and a team of *Geographic* photographers
visited a village in Ceylon to witness the alleged miracles of a local
wonderworker named Mohotty. It seems that as a young boy Mohotty
prayed to a Ceylonese divinity named Kataragama and told the god
that if he cleared Mohotty's father of a murder charge, he, Mohotty,
would do yearly penance in Kataragama's honor. Mohotty's father
was cleared, and true to his word, every year Mohotty did his penance.

This consisted of walking through fire and hot coals, piercing his
cheeks with skewers, driving skewers into his arms from shoulder to
wrist, sinking large hooks deep into his back, and dragging an enormous sledge around a courtyard with ropes attached to the hooks. As
the Grosvenors later reported, the hooks pulled the flesh in Mohotty's
back quite taut, and again there was no sign of blood. When Mohotty
was finished and the hooks were removed, there weren't even any
traces of wounds. The *Geographic* team photographed this unnerving
display and published both pictures and an account of the incident in
the April 1966 issue of *National Geographic*.[58]

In 1967 *Scientific American* published a report about a similar
annual ritual in India. In that instance a *different* person was chosen
each year by the local community, and after a generous amount of
ceremony, two hooks large enough to hang a side of beef on were
buried in the victim's back. Ropes that were pulled through the eyes
of the hooks were tied to the boom of an ox cart, and the victim was
then swung in huge arcs over the fields as a sacramental offering to
the fertility gods. When the hooks were removed the victim was completely unharmed, there was no blood, and literally no sign of any
punctures in the flesh itself.[59]

OUR UNCONSCIOUS BELIEFS

As we have seen, if we are not fortunate enough to have the self-
mastery of a Dajo or a Mohotty, another way of accessing the healing
force within us is to bypass the thick armor of doubt and skepticism

that exists in our conscious minds. Being tricked with a placebo is one way of accomplishing this. Hypnosis is another. Like a surgeon reaching in and altering the condition of an internal organ, a skilled hypnotherapist can reach into our psyche and help us change the most important type of belief of all, our unconscious beliefs.

Numerous studies have demonstrated irrefutably that under hypnosis a person can influence processes usually considered unconscious. For instance, like a multiple, deeply hypnotized persons can control allergic reactions, blood flow patterns, and nearsightedness. In addition, they can control heart rate, pain, body temperature, and even will away some kinds of birthmarks. Hypnosis can also be used to accomplish something that, in its own way, is every bit as remarkable as suffering no injury after a foil has been stuck through one's abdomen.

That something involves a horribly disfiguring hereditary condition known as Brocq's disease. Victims of Brocq's disease develop a thick, horny covering over their skin that resembles the scales of a reptile. The skin can become so hardened and rigid that even the slightest movement will cause it to crack and bleed. Many of the so-called alligator-skinned people in circus sideshows were actually individuals with Brocq's disease, and because of the risk of infection, victims of Brocq's disease used to have relatively short lifespans.

Brocq's disease was incurable until 1951 when a sixteen-year-old boy with an advanced case of the affliction was referred as a last resort to a hypnotherapist named A. A. Mason at the Queen Victoria Hospital in London. Mason discovered that the boy was a good hypnotic subject and could easily be put into a deep state of trance. While the boy was in trance, Mason told him that his Brocq's disease was healing and would soon be gone. Five days later the scaly layer covering the boy's left arm fell off, revealing soft, healthy flesh beneath. By the end of ten days the arm was completely normal. Mason and the boy continued to work on different body areas until all of the scaly skin was gone. The boy remained symptom-free for at least five years, at which point Mason lost touch with him.[60]

This is extraordinary because Brocq's disease is a genetic condition, and getting rid of it involves more than just controlling autonomic processes such as blood flow patterns and various cells of the immune system. It means tapping into the masterplan, our DNA programming itself. So, it would appear that when we access the right strata of our beliefs, our minds can override even our genetic makeup.

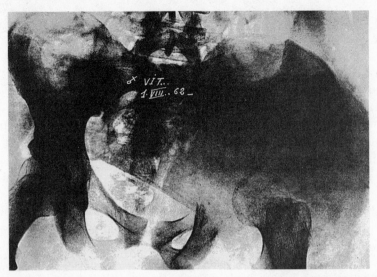

FIGURE 11. A 1962 X ray showing the degree to which Vittorio Michelli's hip bone had disintegrated as a result of his malignant sarcoma. So little bone was left that the ball of his upper leg was free-floating in a mass of soft tissue, rendered as gray mist in the X ray.

FIGURE 12. After a series of baths in the spring at Lourdes, Michelli experienced a miraculous healing. His hip bone completely regenerated over the course of several months, a feat currently considered impossible by medical science. This 1965 X ray shows his miraculously restored hip joint. [Source: Michel-Marie Salmon, *The Extraordinary Cure of Vittorio Michelli*. Used by permission]

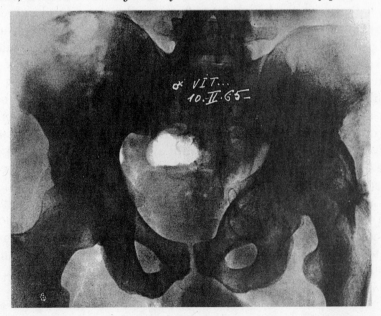

THE BELIEFS EMBODIED IN OUR FAITH

Perhaps the most powerful types of belief of all are those we express through spiritual faith. In 1962 a man named Vittorio Michelli was admitted to the Military Hospital of Verona, Italy, with a large cancerous tumor on his left hip (see fig. 11). So dire was his prognosis that he was sent home without treatment, and within ten months his hip had completely disintegrated, leaving the bone of his upper leg floating in nothing more than a mass of soft tissue. He was, quite literally, falling apart. As a last resort he traveled to Lourdes and had himself bathed in the spring (by this time he was in a plaster cast, and his movements were quite restricted). Immediately on entering the water he had a sensation of heat moving through his body. After the bath his appetite returned and he felt renewed energy. He had several more baths and then returned home.

Over the course of the next month he felt such an increasing sense of well-being he insisted his doctors X-ray him again. They discovered his tumor was smaller. They were so intrigued they documented every step in this improvement. It was a good thing because after Michelli's tumor disappeared, his bone began to regenerate, and the medical community generally views this as an impossibility. Within two months he was up and walking again, and over the course of the next several years his bone completely reconstructed itself (see fig. 12).

A dossier on Michelli's case was sent to the Vatican's Medical Commission, an international panel of doctors set up to investigate such matters, and after examining the evidence the commission decided Michelli had indeed experienced a miracle. As the commission stated in its official report, "A remarkable reconstruction of the iliac bone and cavity has taken place. The X rays made in 1964, 1965, 1968 and 1969 confirm categorically and without doubt that an unforeseen and even overwhelming bone reconstruction has taken place of a type unknown in the annals of world medicine."*[61]

Was Michelli's healing a miracle in the sense that it violated any of the known laws of physics? Although the jury remains out on this question, there seems no clear-cut reason to believe any laws were

*In a truly stunning example of synchronicity, while I was in the middle of writing these very words a letter arrived in the mail informing me that a friend who lives in Kauai, Hawaii, and whose hip had disintegrated due to cancer has also experienced an "inexplicable" and complete regeneration of her bone. The tools she employed to effect her recovery were chemotherapy, extensive meditation, and imagery exercises. The story of her healing has been reported in the Hawaiian newspapers.

violated. Rather, Michelli's healing may simply be due to natural processes we do not yet understand. Given the phenomenal range of
healing capacities we have looked at so far, it is clear there are many
pathways of interaction between the mind and body that we do not yet
understand.

If Michelli's healing was attributable to an undiscovered natural
process, we might better ask, Why is the regeneration of bone so rare
and what triggered it in Michelli's case? It may be that bone regeneration is rare because achieving it requires the accessing of very deep
levels of the psyche, levels usually not reached through the normal
activities of consciousness. This appears to be why hypnosis is needed
to bring about a remission of Brocq's disease. As for what triggered
Michelli's healing, given the role belief plays in so many examples of
mind/body plasticity it is certainly a primary suspect. Could it be that
through his faith in the healing power of Lourdes, Michelli somehow,
either consciously or serendipitously, effected his own cure?

There is strong evidence that belief, not divine intervention, is the
prime mover in at least some so-called miraculous occurrences. Recall
that Mohotty attained his supernormal self-control by praying to Kataragama, and unless we are willing to accept the existence of Kataragama, Mohotty's abilities seem better explained by his deep and abiding *belief* that he was divinely protected. The same seems to be true of
many miracles produced by Christian wonder-workers and saints.

One Christian miracle that appears to be generated by the power of
the mind is stigmata. Most church scholars agree that St. Francis of
Assisi was the first person to manifest spontaneously the wounds of
the crucifixion, but since his death there have been literally hundreds
of other stigmatists. Although no two ascetics exhibit the stigmata in
quite the same way, all have one thing in common. From St. Francis
on, all have had wounds on their hands and feet that represent where
Christ was nailed to the cross. This is not what one would expect if
stigmata were God-given. As parapsychologist D. Scott Rogo, a member of the graduate faculty at John F. Kennedy University in Orinda,
California, points out, it was Roman custom to place the nails through
the *wrists*, and skeletal remains from the time of Christ bear this out.
Nails inserted through the hands cannot support the weight of a body
hanging on a cross.[62]

Why did St. Francis and all the other stigmatists who came after him
believe the nail holes passed through the hands? Because that is the
way the wounds have been depicted by artists since the eighth cen-

tury. That the position and even size and shape of stigmata have been influenced by art is especially apparent in the case of an Italian stigmatist named Gemma Galgani, who died in 1903. Gemma's wounds precisely mirrored the stigmata on her own favorite crucifix.

Another researcher who believed stigmata are self-induced was Herbert Thurston, an English priest who wrote several volumes on miracles. In his tour de force *The Physical Phenomena of Mysticism*, published posthumously in 1952, he listed several reasons why he thought stigmata were a product of autosuggestion. The size, shape, and location of the wounds varies from stigmatist to stigmatist, an inconsistency that indicates they are not derived from a common source, i.e., the actual wounds of Christ. A comparison of the visions experienced by various stigmatists also shows little consistency, suggesting that they are not reenactments of the historical crucifixion, but are instead products of the stigmatists' own minds. And perhaps most significant of all, a surprisingly large percentage of stigmatists also suffered from hysteria, a fact Thurston interpreted as a further indication that stigmata are the side effect of a volatile and abnormally emotional psyche, and not necessarily the product of an enlightened one.[63] In view of such evidence it is small wonder that even some of the more liberal members of the Catholic leadership believe stigmata are the product of "mystical contemplation," that is, that they are *created* by the mind during periods of intense meditation.

If stigmata are products of autosuggestion, the range of control the mind has over the body holographic must be expanded even further. Like Mohotty's wounds, stigmata can also heal with disconcerting speed. The almost limitless plasticity of the body is further evidenced in the ability of some stigmatists to grow nail-like protuberances in the middle of their wounds. Again, St. Francis was the first to display this phenomenon. According to Thomas of Celano, an eyewitness to St. Francis's stigmata and also his biographer: "His hands and feet seemed pierced in the midst by nails. These marks were round on the inner side of the hands and elongated on the outer side, and certain small pieces of flesh were seen like the ends of nails bent and driven back, projecting from the rest of the flesh."[64]

Another contemporary of St. Francis's, St. Bonaventura, also witnessed the saint's stigmata and said that the nails were so clearly defined one could slip a finger under them and into the wounds. Although St. Francis's nails appeared to be composed of blackened and hardened flesh, they possessed another naillike quality. According to

Thomas of Celano, if a nail were pressed on one side, it instantly projected on the other side, just as it would if it were a real nail being slid back and forth through the middle of the hand!

Therese Neumann, the well-known Bavarian stigmatist who died in 1962, also had such naillike protuberances. Like St. Francis's they were apparently formed of hardened skin. They were thoroughly examined by several doctors and found to be structures that passed completely through her hands and feet. Unlike St. Francis's wounds, which were open continuously, Neumann's opened only periodically, and when they stopped bleeding, a soft, membranelike tissue quickly grew over them.

Other stigmatists have displayed similarly profound alterations in their bodies. Padre Pio, the famous Italian stigmatist who died in 1968, had stigmata wounds that passed completely through his hands. A wound in his side was so deep that doctors who examined it were afraid to measure it for fear of damaging his internal organs. Venerable Giovanna Maria Solimani, an eighteenth-century Italian stigmatist, had wounds in her hands deep enough to stick a key into. As with all stigmatists' wounds, hers never became decayed, infected, or even inflamed. And another eighteenth-century stigmatist, St. Veronica Giuliani, an abbess at a convent in Citta di Castello in Umbria, Italy, had a large wound in her side that *would open and close on command*.

Images Projected Outside the Brain

The holographic model has aroused the interest of researchers in the Soviet Union, and two Soviet psychologists, Dr. Alexander P. Dubrov and Dr. Veniamin N. Pushkin, have written extensively on the idea. They believe that the frequency processing capabilities of the brain do not in and of themselves prove the holographic nature of the images and thoughts in the human mind. They have, however, suggested what might constitute such proof. Dubrov and Pushkin believe that if an example could be found where the brain projected an image outside of itself, the holographic nature of the mind would be convincingly demonstrated. Or to use their own words, "Records of ejection of psychophysical structures outside the brain would provide direct evidence of brain holograms."[65]

In fact, St. Veronica Giuliani seems to supply such evidence. During the last years of her life she became convinced that the images of the Passion—a crown of thorns, three nails, a cross, and a sword—had become emblazoned on her heart. She drew pictures of these and even noted where they were located. After she died an autopsy revealed that the symbols were indeed impressed on her heart exactly as she had depicted them. The two doctors who performed the autopsy signed sworn statements attesting to their finding.[66]

Other stigmatists have had similar experiences. St. Teresa of Avila had a vision of an angel piercing her heart with a sword, and after she died a deep fissure was found in her heart. Her heart, with the miraculous sword wound still clearly visible, is now on display as a relic in Alba de Tormes, Spain.[67] A nineteenth-century French stigmatist named Marie-Julie Jahenny kept seeing the image of a flower in her mind, and eventually a picture of the flower appeared on her breast. It remained there twenty years.[68] Nor are such abilities limited to stigmatists. In 1913 a twelve-year-old girl from the village of Bussus-Bus-Suel, near Abbeville, France, made headlines when it was discovered that she could consciously command images, such as pictures of dogs and horses, to appear on her arms, legs, and shoulders. She could also produce words, and when someone asked her a question the answer would instantly appear on her skin.[69]

Surely such demonstrations are examples of the ejection of psychophysical structures outside the brain. In fact, in a way stigmata themselves, especially those in which the flesh has formed into nail-like protrusions, are examples of the brain projecting images outside itself and impressing them in the soft clay of the body holographic. Dr. Michael Grosso, a philosopher at Jersey City State College who has written extensively on the subject of miracles, has also arrived at this conclusion. Grosso, who traveled to Italy to study Padre Pio's stigmata firsthand, states, "One of the categories in my attempt to analyze Padre Pio is to say that he had an ability to symbolically transform physical reality. In other words, the level of consciousness he was operating at enabled him to transform physical reality in the light of certain symbolic ideas. For example, he identified with the wounds of the crucifixion and his body became permeable to those psychic symbols, gradually assuming their form."[70]

So it appears that through the use of images, the brain can tell the body what to do, including telling it to make more images. Images making images. Two mirrors reflecting each other infinitely. Such is the nature of the mind/body relationship in a holographic universe.

Laws Both Known and Unknown

At the beginning of this chapter, I said that instead of examining the various mechanisms the mind uses to control the body, the chapter would be devoted primarily to exploring the range of this control. In doing so I did not mean to deny or diminish the importance of such mechanisms. They are crucial to our understanding of the mind/body relationship, and new discoveries in this area seem to appear every day.

For example, at a recent conference on psychoneuroimmunology—a new science that studies the way the mind (psycho), the nervous system (neuro), and the immune system (immunology) interact—Candace Pert, chief of brain biochemistry at the National Institute of Mental Health, announced that immune cells have neuropeptide receptors. Neuropeptides are molecules the brain uses to communicate, the brain's telegrams, if you will. There was a time when it was believed that neuropeptides could only be found in the brain. But the existence of receptors (telegram receivers) on the cells in our immune system implies that the immune system is not separate from but is an extension of the brain. Neuropeptides have also been found in various other parts of the body, leading Pert to admit that she can no longer tell where the brain leaves off and the body begins.[71]

I have excluded such particulars, not only because I felt examining the extent to which the mind can shape and control the body was more relevant to the discussion at hand, but also because the biological processes responsible for mind/body interactions are too vast a subject for this book. At the beginning of the section on miracles I said there was no clear-cut reason to believe Michelli's bone regeneration could not be explained by our current understanding of physics. This is less true of stigmata. It also appears to be very much not true of various paranormal phenomena reported by credible individuals throughout history, and in recent times by various biologists, physicists, and other researchers.

In this chapter we have looked at astounding things the mind can do that, although not fully understood, do not seem to violate any of the known laws of physics. In the next chapter we will look at some of the things the mind can do that cannot be explained by our current scientific understandings. As we will see, the holographic idea may shed light in these areas as well. Venturing into these territories will occasionally involve treading on what might at first seem to be shaky ground and examining phenomena even more dizzying and incredible than Mohotty's rapidly healing wounds and the images on St. Veronica Giuliani's heart. But again we will find that, despite their daunting nature, science is also beginning to make inroads into these territories.

Acupuncture Microsystems and the Little Man in the Ear

Before closing, one last piece of evidence of the body's holographic nature deserves to be mentioned. The ancient Chinese art of acupuncture is based on the idea that every organ and bone in the body is connected to specific points on the body's surface. By activating these acupuncture points, with either needles or some other form of stimulation, it is believed that diseases and imbalances affecting the parts of the body connected to the points can be alleviated and even cured. There are over a thousand acupuncture points organized in imaginary lines called meridians on the body's surface. Although still controversial, acupuncture is gaining acceptance in the medical community and has even been used successfully to treat chronic back pain in racehorses.

In 1957 a French physician and acupuncturist named Paul Nogier published a book called *Treatise of Auriculotherapy*, in which he announced his discovery that in addition to the major acupuncture system, there are two smaller acupuncture systems on both ears. He dubbed these *acupuncture microsystems* and noted that when one played a kind of connect-the-dots game with them, they formed an anatomical map of a miniature human inverted like a fetus (see fig. 13). Unbeknownst to Nogier, the Chinese had discovered the "little man in the ear" nearly 4,000 years earlier, but a map of the Chinese ear system wasn't published until after Nogier had already laid claim to the idea.

The little man in the ear is not just a charming aside in the history of acupuncture. Dr. Terry Oleson, a psychobiologist at the Pain Management Clinic at the University of California at Los Angeles School of Medicine, has discovered that the ear microsystem can be used to diagnose accurately what's going on in the body. For instance, Oleson has discovered that increased electrical activity in one of the acupuncture points in the ear generally indicates a pathological condition (either past or present) in the corresponding area of the body. In one study, forty patients were examined to determine areas of their body where they experienced chronic pain. Following the examination, each

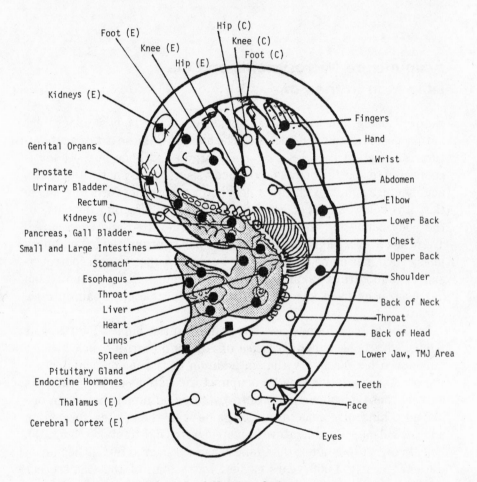

C = Chinese Ear Acupuncture System

E = European Auriculotherapy System

FIGURE 13. The Little Man in the Ear. Acupuncturists have found that the acupuncture points in the ear form the outline of a miniature human being. Dr. Terry Oleson, a psychobiologist at UCLA's School of Medicine, believes it is because the body is a hologram and each of its portions contains an image of the whole. [Copyright Dr. Terry Oleson, UCLA School of Medicine. Used by permission]

patient was draped in a sheet to conceal any visible problems. Then an acupuncturist with no knowledge of the results examined only their ears. When the results were tallied it was discovered that the ear examinations were in agreement with the established medical diagnoses 75.2 percent of the time.[72]

Ear examinations can also reveal problems with the bones and internal organs. Once when Oleson was out boating with an acquaintance he noticed an abnormally flaky patch of skin in one of the man's ears. From his research Oleson knew the spot corresponded to the heart, and he suggested to the man that he might want to get his heart checked. The man went to his doctor the next day and discovered he had a cardiac problem which required immediate open-heart surgery.[73]

Oleson also uses electrical stimulation of the acupuncture points in the ear to treat chronic pain, weight problems, hearing loss, and virtually all kinds of addiction. In one study of 14 narcotic-addicted individuals, Oleson and his colleagues used ear acupuncture to eliminate the drug requirements of 12 of them in an average of 5 days and with only minimal withdrawal symptoms.[74] Indeed, ear acupuncture has proved so successful in bringing about rapid narcotic detoxification that clinics in both Los Angeles and New York are now using the technique to treat street addicts.

Why would the acupuncture points in the ear be aligned in the shape of a miniature human? Oleson believes it is because of the holographic nature of the mind and body. Just as every portion of a hologram contains the image of the whole, every portion of the body may also contain the image of the whole. "The ear holograph is, logically, connected to the brain holograph which itself is connected to the whole body," he states. "The way we use the ear to affect the rest of the body is by working through the brain holograph."[75]

Oleson believes there are probably acupuncture microsystems in other parts of the body as well. Dr. Ralph Alan Dale, the director of the Acupuncture Education Center in North Miami Beach, Florida, agrees. After spending the last two decades tracking down clinical and research data from China, Japan, and Germany, he has accumulated evidence of eighteen different microacupuncture holograms in the body, including ones in the hands, feet, arms, neck, tongue, and even the gums. Like Oleson, Dale feels these microsystems are "holographic reiterations of the gross anatomy," and believes there are still other such systems waiting to be discovered. In a notion reminiscent of Bohm's assertion that every electron in some way contains the cosmos, Dale hypothesizes that every finger, and even every cell, may contain its own acupuncture microsystem.[76]

Richard Leviton, a contributing editor at *East West* magazine, who has written about the holographic implications of acupuncture microsystems, thinks that alternative medical techniques—such as reflexology, a type of massage therapy that involves accessing all points of the body through stimulation of the feet, and iridology, a diagnostic technique that involves examining the iris of the eye in order to determine the condition of the body—may also be indications of the body's holographic nature. Leviton concedes that neither field has been experimentally vindicated (studies of iridology, in particular, have produced extremely conflicting results) but feels the holographic idea offers a way of understanding them if their legitimacy is established.

Leviton thinks there may even be something to palmistry. By this he does not mean the type of hand reading practiced by fortune-tellers who sit in glass storefronts and beckon people in, but the 4,500-year-old Indian version of the science. He bases this suggestion on his own profound encounter with an Indian hand reader living in Montreal who possessed a doctorate in the subject from Agra University, India. "The holographic paradigm provides palmistry's more esoteric and controversial claims a context for validation," says Leviton.[77]

It is difficult to assess the type of palmistry practiced by Leviton's Indian hand reader in the absence of double-blind studies, but science is beginning to accept that at least some information about our body is contained in the lines and whorls of our hand. Herman Weinreb, a neurologist at New York University, has discovered that a fingerprint pattern called an *ulnar loop* occurs more frequently in Alzheimer's

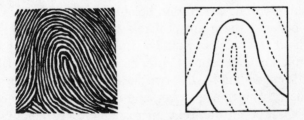

FIGURE 14. Neurologists have found that Alzheimer's patients have a more than average chance of having a distinctive fingerprint pattern known as an *ulnar loop*. At least ten other common genetic disabilities are also associated with various patterns in the hand. Such findings may provide evidence of the holographic model's assertion that every portion of the body contains information about the whole. [Redrawn by the author from original art in *Medicine* magazine]

patients than in nonsufferers (see fig. 14). In a study of 50 Alzheimer's patients and 50 normal individuals, 72 percent of the Alzheimer's group had the pattern on at least 8 of their fingertips, compared to only 26 percent in the control group. Of those with ulnar loops on all 10 fingertips, 14 were Alzheimer's sufferers, but only 4 members of the control group had the pattern.[78]

It is now known that 10 common genetic disabilities, including Down's syndrome, are also associated with various patterns in the hand. Doctors in West Germany are now using this information to analyze parents' hand prints and help determine whether expectant mothers should undergo amniocentesis, a potentially dangerous genetic screening procedure in which a needle is inserted into the womb to draw off amniotic fluid for laboratory testing.

Researchers at West Germany's Institute of Dermatoglyphics in Hamburg have even developed a computer system that uses an optoelectric scanner to take a digitized "photo" of a patient's hand. It then compares the hand to the 10,000 other prints in its memory, scans it for the nearly 50 distinctive patterns now known to be associated with various hereditary disabilities, and quickly calculates the patient's risk factors.[79] So perhaps we should not be so quick to dismiss palmistry out of hand. The lines and whorls in our palms may contain more about our whole self than we realize.

Harnessing the Powers of the Holographic Brain

Throughout this chapter two broad messages come through loud and clear. According to the holographic model, the mind/body ultimately cannot distinguish the difference between the neural holograms the brain uses to experience reality and the ones it conjures up while imagining reality. Both have a dramatic effect on the human organism, an effect so powerful that it can modulate the immune system, duplicate and/or negate the effects of potent drugs, heal wounds with amazing rapidity, melt tumors, override our genetic programming, and reshape our living flesh in ways that almost defy belief. This then is the first message: that each of us possesses the ability, at least at some level, to influence our health and control our physical form in ways that are nothing short of dazzling. We are all potential wonderworkers, dormant yogis, and it is clear from the evidence presented in the preceding pages that it would behoove us both as individuals and as a species to devote a good deal more effort into exploring and harnessing these talents.

The second message is that elements that go into the making of these neural holograms are many and subtle. They include the images upon which we meditate, our hopes and fears, the attitudes of our doctors, our unconscious prejudices, our individual and cultural beliefs, and our faith in things both spiritual and technological. More than just facts, these are important clues, signposts that point toward those things that we must become aware of and acquire mastery over if we are to learn how to unleash and manipulate these talents. There are, no doubt, other factors involved, other influences that shape and circumscribe these abilities, for one thing should now be obvious. In a holographic universe, a universe in which a slight change in attitude can mean the difference between life and death, in which things are so subtly interconnected that a dream can call forth the inexplicable appearance of a scarab beetle, and the factors responsible for an illness can also evoke a certain pattern in the lines and whorls of the hand, we have reason to suspect that each effect has multitudinous causes. Each linkage is the starting point of a dozen more, for in the words of Walt Whitman, "A vast similitude interlocks all."

5

A Pocketful of Miracles

Miracles happen, not in opposition to Nature,
but in opposition to what we know of Nature.

—St. Augustine

Every year in September and May a huge crowd gathers at the Duomo San Gennaro, the principal cathedral of Naples, to witness a miracle. The miracle involves a small vial containing a brown crusty substance alleged to be the blood of San Gennaro, or St. Januarius, who was beheaded by the Roman emperor Diocletian in A.D. 305. According to legend, after the saint was martyred a serving woman collected some of his blood as a relic. No one knows precisely what happened after that, save that the blood didn't turn up again until the end of the thirteenth century when it was ensconced in a silver reliquary in the cathedral.

The miracle is that twice yearly, when the crowd shouts at the vial, the brown crusty substance changes into a bubbling, bright red liquid. There is little doubt that the liquid is real blood. In 1902 a group of scientists from the University of Naples made a spectroscopic analysis of the liquid by passing a beam of light through it, verifying that it was blood. Unfortunately, because the reliquary containing the blood is so old and fragile, the church will not allow it to be cracked open so that other tests can be done, and so the phenomenon has never been thoroughly studied.

But there is further evidence that the transformation is a more than ordinary event. Occasionally throughout history (the first written account of the public performance of the miracle dates back to 1389) when the vial is brought out, the blood refuses to liquefy. Although rare, this is considered a very bad omen by the citizens of Naples. In the past, the failure of the miracle has directly preceded the eruption of Vesuvius and the Napoleonic invasion of Naples. More recently, in 1976 and 1978, it presaged the worst earthquake in Italian history and the election of a communist city government in Naples, respectively.

Is the liquefaction of San Gennaro's blood a miracle? It appears to be, at least in the sense that it seems impossible to explain by known scientific laws. Is the liquefaction caused by San Gennaro himself? My own feeling is that its more likely cause is the intense devotion and belief of the people witnessing the miracle. I say this because nearly all of the miracles performed by saints and wonder-workers of the world's great religions have also been duplicated by psychics. This suggests that, as with stigmata, miracles are produced by forces lying deep in the human mind, forces that are latent in all of us. Herbert Thurston, the priest who wrote *The Physical Phenomena of Mysticism*, himself was aware of this similarity and was reluctant to attribute any miracle to a truly supernatural cause (as opposed to a psychic or paranormal cause). Another piece of evidence supportive of this idea is that many stigmatists, including Padre Pio and Therese Neumann, were also renowned for their psychic abilities.

One psychic ability that appears to play a role in miracles is psychokinesis or PK. Since the miracle of San Gennaro involves a physical alteration of matter, PK is certainly a likely suspect. Rogo believes PK is also responsible for some of the more dramatic aspects of stigmata. He feels that it is well within the normal biological capabilities of the body to cause small blood vessels under the skin to break and produce superficial bleeding, but only PK can account for the rapid appearance of large wounds.[1] Whether this is true or not remains to be seen, but PK is clearly a factor in some of the phenomena that accompany stigmata. When blood flowed from the wounds in Therese Neumann's feet, it always flowed toward her toes—exactly as it would have flowed from Christ's wounds when he was on the cross—regardless of how her feet were positioned. This meant that when she was sitting upright in bed, the blood actually flowed *upward and counter to the force of gravity.* This was observed by numerous witnesses, including many U.S. servicemen stationed in Germany after the war who visited Neumann to witness her miraculous abilities. Gravity-defying flows of blood have been reported in other cases of stigmata as well.[2]

Such events leave us agog because our current worldview does not provide us with a context with which to understand PK. Bohm believes viewing the universe as a holomovement does provide us with a context. To explain what he means he asks us to consider the following situation. Imagine you are walking down a street late one night and a shadow suddenly looms up out of nowhere. Your first thought might be that the shadow is an assailant and you are in danger. The information contained in this thought will in turn give rise to a range of imagined activities, such as running, being hurt, and fighting. The presence of these imagined activities in your mind, however, is *not* a purely "mental" process, for they are inseparable from a host of related biological processes, such as excitation of nerves, rapid heart beat, release of adrenaline and other hormones, tensing of the muscles, and so on. Conversely, if your first thought is that the shadow is just a shadow, a different set of mental and biological responses will follow. Moreover, a little reflection will reveal that we react both mentally and biologically to everything we experience.

According to Bohm, the important point to be gleaned from this is that consciousness is not the only thing that can respond to *meaning.* The body can also respond, and this reveals that meaning is simultaneously both mental and physical in nature. This is odd, for we normally think of meaning as something that can only have an active effect on subjective reality, on the thoughts inside our heads, not something that can engender a response in the physical world of things and objects. Meaning "can thus serve as the link or 'bridge' between these two sides of reality," Bohm states. "This link is indivisible in the sense that information contained in thought, which we feel to be on the 'mental' side, is at the same time a neurophysiological, chemical, and physical activity, which is clearly what is meant by this thought on the 'material' side."[3]

Bohm feels that examples of objectively active meaning can be found in other physical processes. One is the functioning of a computer chip. A computer chip contains information, and the meaning of the information is active in the sense that it determines how electrical currents flow through the computer. Another is the behavior of subatomic particles. The orthodox view in physics is that quantum waves act mechanically on a particle, controlling its movement in much the same way that the waves of the ocean might control a Ping-Pong ball floating on its surface. But Bohm does not feel that this view can explain, for example, the coordinated dance of electrons in a plasma any more than the wave motion of water could explain a similarly well-choreographed movement of Ping-Pong balls if such a movement were discovered on the ocean's surface. He believes the relationship between particle and quantum wave is more like a ship on automatic pilot guided by radar waves. A quantum wave does not push an electron about any more than a radar wave pushes a ship. Rather, it provides the electron with *information* about its environment which the electron then uses to maneuver on its own.

In other words, Bohm believes that an electron is not only mindlike, but is a highly complex entity, a far cry from the standard view that an electron is a simple, structureless point. The active use of information by electrons, and indeed by all subatomic particles, indicates that the ability to respond to meaning is a characteristic not only of consciousness but of all matter. It is this intrinsic commonality, says Bohm, that offers a possible explanation for PK. He states, "On this basis, psychokinesis could arise if the mental processes of one or more people were focused on meanings that were in harmony with those guiding the basic processes of the material systems in which this psychokinesis was to be brought about."[4]

It is important to note that this kind of psychokinesis would not be due to a causal process, that is, a cause-and-effect relationship involving any of the known forces in physics. Instead, it would be the result of a kind of nonlocal "resonance of meanings," or a kind of nonlocal interaction similar to, but not the same as, the nonlocal interconnection that allows a pair of twin photons to manifest the same angle of polarization which we saw in chapter 2 (for technical reasons Bohm believes mere quantum nonlocality cannot account for either PK or telepathy, and only a deeper form of nonlocality, a kind of "super" nonlocality, would offer such an explanation).

The Gremlin in the Machine

Another researcher whose ideas about PK are similar to Bohm's, but who has taken them one step further, is Robert G. Jahn, a professor of aerospace sciences and dean emeritus of the School of Engineering and Applied Science at Princeton University. Jahn's involvement in the study of PK happened quite by accident. A former consultant for both NASA and the Department of Defense, his original field of interest was deep space propulsion. In fact, he is the author of *Physics of Electric Propulsion,* the leading textbook in the field, and didn't even believe in the paranormal when a student first approached him and asked him to oversee a PK experiment she wanted to do as an independent study project. Jahn reluctantly agreed, and the results were so provocative they inspired him to found the Princeton Engineering Anomalies Research (PEAR) lab in 1979. Since then PEAR researchers have not only produced compelling evidence of the existence of PK, but have gathered more data on the subject than anyone else in the country.

In one series of experiments Jahn and his associate, clinical psychologist Brenda Dunne, employed a device called a random event generator, or REG. By relying on an unpredictable natural process such as radioactive decay, a REG is able to produce a string of random binary numbers. Such a string might look something like this: 1, 2, 1, 2, 2, 1, 1, 2, 1, 1, 1, 2, 1. In other words, a REG is a kind of automatic coin-flipper capable of producing an enormous number of coin flips in a very short time. As everyone knows, if you flip a perfectly weighted coin 1,000 times, the odds are you will get a 50/50 split between heads and tails. In reality, out of any 1,000 such flips, the split may vary a little in one direction or the other, but the greater the number of flips, the closer to 50/50 the split will become.

What Jahn and Dunne did was have volunteers sit in front of the REG and concentrate on having it produce an abnormally large number of either heads or tails. Over the course of literally hundreds of thousands of trials they discovered that, through concentration alone, the volunteers did indeed have a small but statistically significant effect on the REG's output. They discovered two other things as well. The ability to produce PK effects was not limited to a few gifted individuals but was present in the majority of volunteers they tested. This suggests that most of us possess some degree of PK. They also discovered that different volunteers produced different and consistently distinctive results, results that were so idiosyncratic that Jahn and Dunne started calling them "signatures."[5]

In another series of experiments Jahn and Dunne employed a pinball-like device that allows 9,000 three-quarter-inch marbles to cir-

culate around 330 nylon pegs and distribute themselves into 19 collect-
ing bins at the bottom. The device is contained in a shallow vertical
frame ten feet high and six feet wide with a clear glass front so that
volunteers can see the marbles as they fall and collect in the bins.
Normally, more balls fall in the center bins than in the outer ones, and
the overall distribution looks like a bell-shaped curve.

As with the REG, Jahn and Dunne had volunteers sit in front of the
machine and try to make more balls land in the outer bins than in the
center ones. Again, over the course of a large number of runs, the
operators were able to create a small but measurable shift in where
the balls landed. In the REG experiments the volunteers only exerted
a PK effect on microscopic processes, the decay of a radioactive substance, but the pinball experiments revealed that test subjects could
use PK to influence objects in the everyday world as well. What's
more, the "signatures" of individuals who had participated in the REG
experiments surfaced again in the pinball experiments, suggesting
that the PK abilities of any given individual remain the same from
experiment to experiment, but vary from individual to individual just
as other talents vary. Jahn and Dunne state, "While small segments
of these results might reasonably be discounted as falling too close to
chance behavior to justify revision of prevailing scientific tenets, taken
in concert the entire ensemble establishes an incontrovertible aberration of substantial proportions."[6]

Jahn and Dunne think their findings may explain the propensity
some individuals seem to have for jinxing machinery and causing
equipment to malfunction. One such individual was physicist Wolfgang Pauli, whose talents in this area are so legendary that physicists
have jokingly dubbed it the "Pauli effect." It is said that Pauli's mere
presence in a laboratory would cause a glass apparatus to explode, or
a sensitive measuring device to crack in half. In one particularly famous incident a physicist wrote Pauli to say that at least he couldn't
blame Pauli for the recent and mysterious disintegration of a compli-
cated piece of equipment since Pauli had not been present, only to find
that Pauli had been passing by the laboratory in a train at the precise
moment of the mishap! Jahn and Dunne think the famous "Gremlin
effect," the tendency of carefully tested pieces of equipment to undergo inexplicable malfunctions at the most absurdly inopportune moments, often reported by pilots, aircrew, and military operators, may
also be an example of unconscious PK activity.

If our minds can reach out and alter the movement of a cascade of marbles or the operation of a machine, what strange alchemy might account for such an ability? Jahn and Dunne believe that since all known physical processes possess a wave/particle duality, it is not unreasonable to assume that consciousness does as well. When it is particlelike, consciousness would appear to be localized in our heads, but in its wavelike aspect, consciousness, like all wave phenomena, could also produce remote influence effects. They believe one of these remote influence effects is PK.

But Jahn and Dunne do not stop here. They believe that reality is itself the result of the interface between the wavelike aspects of consciousness and the wave patterns of matter. However, like Bohm, they do not believe that consciousness or the material world can be productively represented in isolation, or even that PK can be thought of as the transmission of some kind of force. "The message may be more subtle than that," says Jahn. "It may be that such concepts are simply unviable, that we cannot talk profitably about an abstract environment or an abstract consciousness. The only thing we can experience is the interpenetration of the two in some way."[7]

If PK cannot be thought of as the transmission of some kind of force, what terminology might better sum up the interaction of mind and matter? In thinking that is again similar to Bohm's, Jahn and Dunne propose that PK actually involves an exchange of information between consciousness and physical reality, an exchange that should be thought of less as a flow between the mental and the material, and more as a *resonance* between the two. The importance of resonance was even sensed and commented on by the volunteers in the PK experiments, in that the most frequently mentioned factor associated with a successful performance was the attainment of a feeling of "resonance" with the machine. One volunteer described the feeling as "a state of immersion in the process which leads to a loss of awareness of myself. I don't feel any direct control over the device, more like a marginal influence when I'm in resonance with the machine. It's like being in a canoe; when it goes where I want, I flow with it. When it doesn't I try to break the flow and give it a chance to get back in resonance with me."[8]

Jahn and Dunne's ideas are similar to Bohm's in several other key ways. Like Bohm, they believe that the concepts we use to describe reality—electron, wavelength, consciousness, time, frequency—are useful only as "information-organizing categories" and possess no independent status. They also believe that all theories, including their own, are only metaphors. Although they do not identify themselves with the holographic model (and their theory does in fact differ from Bohm's thinking in several significant ways), they do recognize the overlap. "To the extent that we're talking about a rather basic reliance on wave mechanical behavior, there is some commonality between what we're postulating and the holographic idea," says Jahn. "It gives to consciousness the capacity to function in a wave mechanical sense and thereby to avail itself, one way or another, of all of space and time."[9]

Dunne agrees: "In some sense the holographic model could be perceived as addressing the mechanism whereby the consciousness interacts with that wave mechanical, aboriginal, sensible muchness, and somehow manages to convert it into usable information. In another sense, if you imagine that the individual consciousness has its own characteristic wave patterns, you could view it—metaphorically, of course—as the laser of a particular frequency that intersects with a specific pattern in the cosmic hologram."[10]

As might be expected, Jahn and Dunne's work has been greeted with considerable resistance by the scientific orthodox community, but it is gaining acceptance in some quarters. A good deal of PEAR's funding comes from the McDonnell Foundation, created by James S. McDonnell III, of the McDonnell Douglas Corporation, and the *New York Times Magazine* recently devoted an article to Jahn and Dunne's work. Jahn and Dunne themselves remain undaunted by the fact that they are devoting so much time and effort to exploring the parameters of a phenomenon considered nonexistent by most other scientists. As Jahn states, "My sense of the importance of this topic is much higher than anything else I've ever worked on."[11]

Psychokinesis on a Grander Scale

So far, PK effects produced in the lab have been limited to relatively small objects, but the evidence suggests that some individuals at least can use PK to bring about even greater changes in the physical world. Biologist Lyall Watson, author of the bestselling book *Supernature* and a scientist who has studied paranormal events all over the world, encountered one such individual while visiting the Philippines. The man was one of the so-called Philippine psychic healers, but instead of touching a patient, all he did was hold his hand about ten inches over the person's body, point at his or her skin, and an incision would appear instantaneously. Watson not only witnessed several displays of the man's psychokinetic surgical skills, but once, when the man made a broader sweep with his finger than usual, Watson received an incision on the back of his own hand. He bears the scar to this day.[11]

There is evidence that PK abilities can also be used to heal bones. Several examples of such healings have been reported by Dr. Rex Gardner, a physician at Sunderland District General Hospital in England. One interesting aspect of a 1983 article in the *British Medical Journal* is that Gardner, an avid investigator of miracles, presents contemporary miraculous healings side by side with examples of virtually identical healings collected by seventh-century English historian and theologian the Venerable Bede.

The present-day healing involved a group of Lutheran nuns living in Darmstadt, Germany. The nuns were building a chapel when one of the sisters broke through a freshly cemented floor and fell onto a wooden beam below. She was rushed to the hospital where X rays revealed that she had a compound pelvic fracture. Instead of relying on standard medical techniques, the nuns held an all-night prayer vigil. Despite the doctors' insistence that the sister should remain in traction for many weeks, the nuns took her home two days later and continued to pray and perform a laying on of hands. To their surprise, immediately following the laying on of hands, the sister stood up from her bed, free of the excruciating pain of the fracture and apparently healed. It took her only two weeks to achieve a full recovery, whereupon she returned to the hospital and presented herself to her astonished doctor.[13]

Although Gardner does not try to account for this or any of the other healings he discusses in his article, PK seems a likely explanation. Given that the natural healing of a fracture is a lengthy process, and even the miraculous regeneration of Michelli's pelvis took several months, it is suggested that perhaps the unconscious PK abilities of the nuns performing the laying on of hands accomplished the task.

Gardner describes a similar healing that occurred in the seventh century during the building of the church at Hexham, England, and involving St. Wilfrid, then the bishop of Hexham. During the construction of the church a mason named Bothelm fell from a great height, breaking both his arms and legs. As he lay dying, Wilfrid prayed over him and asked the other workmen to join him. They did, "the breath of life returned" to Bothelm, and he healed rapidly. Since the healing apparently did not take place until St. Wilfred asked the other workmen to join him, one wonders if St. Wilfred was the catalyst, or again if it was the combined unconscious PK of the entire assemblage?

Dr. William Tufts Brigham, the curator of the Bishop Museum in Honolulu and a noted botanist who devoted much of his private life to investigating the paranormal, recorded an incident in which a broken bone was instantaneously healed by a native Hawaiian shaman, or *kahuna*. The incident was witnessed by a friend of Brigham's named J. A. K. Combs. Combs's grandmother-in-law was considered one of the most powerful women kahunas in the islands, and once, while attending a party at the woman's home, Combs observed her abilities firsthand.

On the occasion in question, one of the guests slipped and fell in the beach sand, breaking his leg so severely that the bone ends pressed visibly out against the skin. Recognizing the seriousness of the break, Combs recommended that the man be taken to a hospital immediately, but the elderly kahuna would hear none of it. Kneeling beside the man, she straightened his leg and pushed on the area where the fractured bones pressed out against his skin. After praying and meditating for several minutes she stood up and announced that the healing was finished. The man rose wonderingly to his feet, took a step, and then another. He was completely healed and his leg showed no indication of the break in any way.[14]

Mass Psychokinesis in Eighteenth-Century France

Such incidents notwithstanding, one of the most astounding manifestations of psychokinesis, and one of the most remarkable displays of miraculous events ever recorded, took place in Paris in the first half of the eighteenth century. The events centered around a puritanical sect of Dutch-influenced Catholics known as the Jansenists, and were precipitated by the death of a saintly and revered Jansenist deacon named François de Paris. Although few people living today have even heard of the Jansenist miracles, they were one of the most talked about events in Europe for the better part of a century.

To understand fully the Jansenist miracles, it is necessary to know a little about the historical events that preceded François de Paris's death. Jansenism was founded in the early seventeenth century, and from the start it was at odds with both the Roman Catholic Church and the French monarchy. Many of the beliefs diverged sharply with standard church doctrine but it was a popular movement and quickly gained followers among the French populace. Most damning of all, it was viewed by both the papacy and King Louis XV, a devout Catholic, as Protestantism only masquerading as Catholicism. As a result, both the church and the king were constantly maneuvering to undermine the movement's power. One obstacle to these maneuverings, and one of the factors that contributed to the movement's popularity, was that Jansenist leaders seemed especially skilled at performing miraculous healings. Nonetheless, the church and the monarchy persevered, causing fierce debates to rage throughout France. It was on May 1, 1727, at the height of this power struggle, that François de Paris died and was interred in the parish cemetery of Saint-Médard, Paris.

Because of the abbé's saintly reputation, worshipers began to gather at his tomb, and from the beginning a host of miraculous healings were reported. The ailments thus cured included cancerous tumors, paralysis, deafness, arthritis, rheumatism, ulcerous sores, persistent fevers, prolonged hemorrhaging, and blindness. But this was not all. The mourners also started to experience strange involuntary spasms or convulsions and to undergo the most amazing contortions of their limbs. These seizures quickly proved contagious, spreading like a brush fire until the streets were packed with men, women, and children, all twisting and writhing as if caught up in a surreal enchantment.

It was while they were in this fitful and trancelike state that the "convulsionaires," as they have come to be called, displayed the most phenomenal of their talents. One was the ability to endure without harm an almost unimaginable variety of physical tortures. These included severe beatings, blows from both heavy and sharp objects, and strangulation—*all with no sign of injury, or even the slightest trace of wounds or bruises.*

What makes these miraculous events so unique is that they were witnessed by literally thousands of observers. The frenzied gatherings around Abbé Paris's tomb were by no means short-lived. The cemetery and the streets surrounding it were crowded day and night for years, and even two decades later miracles were still being reported (to give some idea of the enormity of the phenomena, in 1733 it was noted in the public records that over 3,000 volunteers were needed simply to assist the convulsionaires and make sure, for example, that the female participants did not become immodestly exposed during their seizures). As a result, the supernormal abilities of the convulsionaires became an international cause célèbre, and thousands flocked to see them, including individuals from all social strata and officials from every educational, religious, and governmental institution imaginable; numerous accounts, both official and unofficial, of the miracles witnessed are recorded in the documents of the time.

Moreover, many of the witnesses, such as the investigators from the Roman Catholic Church, had a vested interest in refuting the Jansenist miracles, but they still went away confirming them (the Roman Catholic Church later remedied this embarrassing state of affairs by conceding that the miracles existed but were the work of the devil, hence proving that the Jansenists were depraved).

One investigator, a member of the Paris Parliament named Louis-Basile Carre de Montgeron, witnessed enough miracles to fill four thick volumes on the subject, which he published in 1737 under the title *La Verité des Miracles.* In the work he provides numerous examples of the convulsionaries' apparent invulnerability to torture. In one instance a twenty-year-old convulsionaire named Jeanne Maulet leaned against a stone wall while a volunteer from the crowd, "a very strong man," delivered one hundred blows to her stomach with a thirty-pound hammer (the convulsionaires themselves asked to be tortured because they said it relieved the excruciating pain of the convulsions). To test the force of the blows, Montgeron himself then took the hammer and tried it on the stone wall against which the girl had leaned. He wrote, "At the twenty-fifth blow the stone upon which I struck, which had been shaken by the preceding efforts, suddenly became loose and fell on the other side of the wall, making an aperture more than half a foot in size."[15]

Montgeron describes another instance in which a convulsionaire bent back into an arc so that her lower back was supported by "the sharp point of a peg." She then asked that a fifty-pound stone attached to a rope be hoisted to "an extreme height" and allowed to fall with all its weight on her stomach. The stone was hoisted up and allowed to fall again and again, but the woman seemed completely unaffected by it. She effortlessly maintained her awkward position, suffered no pain or harm, and walked away from the ordeal without even so much as a mark on the flesh of her back. Montgeron noted that while the ordeal was in progress she kept crying out, "Strike harder, harder!"[16]

In fact, it appears that nothing could harm the convulsionaires. They could not be hurt by the blows of metal rods, chains, or timbers. The strongest men could not choke them. Some were crucified and afterward showed no trace of wounds.[17] Most mind-boggling of all, they could not even be cut or punctured with knives, swords, or hatchets! Montgeron cites an incident in which the sharpened point of an iron drill was held against the stomach of a convulsionaire and then pounded so violently with a hammer that it seemed "as if it would penetrate through to the spine and rupture all the entrails." But it didn't, and the convulsionaire maintained an "expression of perfect rapture," crying, "Oh, that does me good! Courage, brother; strike twice as hard, if you can!"[18]

Invulnerability was not the only talent the Jansenists displayed during their seizures. Some became clairvoyant and were able to "discern hidden things." Others could read even when their eyes were closed and tightly bandaged, and instances of levitation were reported. One of the levitators, an abbé named Bescherand from Montpellier, was so "forcibly lifted into the air" during his convulsions that even when witnesses tried to hold him down they could not succeed in keeping him from rising up off of the ground.[19]

Although we have all but forgotten about the Jansenist miracles today, they were far from ignored by the intelligentsia of the time. The niece of the mathematician and philosopher Pascal succeeded in having a severe ulcer in her eye vanish within hours as the result of a Jansenist miracle. When King Louis XV tried unsuccessfully to stop the convulsionaires by closing the cemetery of Saint-Médard, Voltaire quipped, "God was forbidden, by order of the King, to work any miracles there." And in his *Philosophical Essays* the Scottish philosopher David Hume wrote, "There surely never was so great a number of miracles ascribed to one person as those which were lately said to have been wrought in France upon the tomb of Abbé Paris. Many of the miracles were immediately proved upon the spot, before judges of unquestioned credit and distinction, in a learned age, and on the most eminent theatre that is now in the world."

How are we to explain the miracles produced by the convulsionaires? Although Bohm is willing to consider the possibility of PK and other paranormal phenomena, he prefers not to speculate about specific events such as the supernormal abilities of the Jansenists. But once again, if we take the testimony of so many witnesses seriously, unless we are willing to concede that God favored the Jansenist Catholics over the Roman, PK seems the likely explanation. That some kind of psychic functioning was involved is strongly suggested by the appearance of other psychic abilities, such as clairvoyance, during the seizures. In addition, we have already looked at a number of examples where intense faith and hysteria have triggered the deeper forces of the mind, and these too were present in ample portions. In fact, instead of being produced by one individual, the psychokinetic effects may have been created by the combined fervor and belief of all those present, and this might account for the unusual vigor of the manifestations. This idea is not new. In the 1920s the great Harvard psychologist William McDougall also suggested that religious miracles might be the result of the collective psychic powers of large numbers of worshipers.

PK would explain many of the convulsionaire's seeming invulnerabilities. In the case of Jeanne Maulet it could be argued that she unconsciously used PK to block the effect of the hammer blows. If the convulsionaires were unconsciously using PK to take control of chains, timbers, and knives, and stop them in their tracks at the precise moment of impact, it would also explain why these objects left no marks or bruises. Similarly, when individuals tried to strangle the Jansenists, perhaps their hands were held in place by PK and although they thought they were squeezing flesh, they were really only flexing in the nothingness.

Reprogramming the Cosmic Motion Picture Projector

PK does not explain every aspect of the convulsionaires' invulnerability, however. There is the problem of inertia—the tendency of an object in motion to stay in motion—to consider. When a fifty-pound stone or a piece of timber comes crashing down, it carries with it a lot of energy, and when it is stopped in its tracks, the energy has to go somewhere. For example, if a person in a suit of armor is struck by a thirty-pound hammer, although the metal of the armor may deflect the blow, the person is still considerably shaken. In the case of Jeanne Maulet it appears that the energy somehow bypassed her body and was transferred to the wall behind her, for as Montgeron noted, the stone was "shaken by the efforts." But in the case of the woman who was arched and had the fifty-pound stone dropped on her abdomen, the matter is less clear. One wonders why she wasn't driven into the ground like a croquet hoop, or why, when they were struck with timbers, the convulsionaires were not knocked off their feet? Where did the deflected energy go?

Again, the holographic view of reality provides a possible answer. As we have seen, Bohm believes that consciousness and matter are just different aspects of the same fundamental something, a something that has its origins in the implicate order. Some researchers believe this suggests that the consciousness may be able to do much more than make a few psychokinetic changes in the material world. For example, Grof believes that if the implicate and explicate orders are an accurate description of reality, "it is conceivable that certain unusual states of consciousness could mediate direct experience of, and intervention in, the implicate order. It would thus be possible to modify phenomena in the phenomenal world by influencing their generative matrix."[20] Put another way, in addition to psychokinetically moving objects around, the mind may also be able to reach down and reprogram the cosmic motion picture projecter that created those objects in the first place. Thus, not only could the conventionally recognized rules of nature, such as inertia, be completely bypassed, but the mind could alter and reshape the material world in ways far more dramatic than even psychokinesis implies.

That this or some other theory must be true is evidenced in another supernormal ability displayed by various individuals throughout history: invulnerability to fire. In his book *The Physical Phenomena of Mysticism*, Thurston gives numerous examples of saints who possessed this ability, one of the most famous being St. Francis of Paula. Not only could St. Francis of Paula hold burning embers in his hands without being harmed, but at his canonization hearings in 1519 eight eyewitnesses testified that they had seen him walk unharmed through the roaring flames of a furnace to repair one of the furnace's broken walls.

The account brings to mind the Old Testament story of Shadrach, Meshach, and Abednego. After capturing Jerusalem, King Nebuchadnezzar ordered everyone to worship a statue of himself. Shadrach, Meshach, and Abednego refused, so Nebuchadnezzar ordered them thrown into a furnace so "exceeding hot" that the flames even burned up the men who threw them in. However, because of their faith, they survived the fire unscathed, and came out with their hair unsinged, their clothing unharmed, and not even the smell of fire upon them. It seems that challenges to faith, such as the one King Louis XV tried to impose on the Jansenists, have engendered miracles in more than one instance.

Although the kahunas of Hawaii do not walk through roaring furnaces, there are reports that they can stroll across hot lava without being harmed. Brigham told of meeting three kahunas who promised to perform the feat for him, and of following them on a lengthy trek to a lava flow near the erupting Kilauea. They chose a 150-foot-wide lava flow that had cooled enough to support their weight, but was so hot that patches of incandescence still coursed through its surface. As Brigham watched, the kahunas took off their sandals and started to recite the lengthy prayers necessary to protect them as they strolled out onto the barely hardened molten rock.

As it turned out, the kahunas had told Brigham earlier that they could confer their fire immunity on him if he wanted to join them, and he had bravely agreed. But as he faced the baking heat of the lava he had second and even third thoughts. "The upshot of the matter was that I sat tight and refused to take off my boots," Brigham wrote in his account of the incident. After they finished invoking the gods, the oldest kahuna scampered out onto the lava and crossed the 150 feet without harm. Impressed, but still adamant about not going, Brigham stood up to watch the next kahuna, only to be given a shove that forced him to break into a run to keep from falling face first onto the incandescent rock.

And run Brigham did. When he reached higher ground on the other side he discovered that one of his boots had burned off and his socks were on fire. But, miraculously, his feet were completely unharmed. The kahunas had also suffered no harm and were rolling in laughter at Brigham's shock. "I laughed too," wrote Brigham. "I was never so relieved in my life as I was to find that I was safe. There is little more that I can tell of this experience. I had a sensation of intense heat on my face and body, but almost no sensation in my feet."[21]

The convulsionaires also occasionally displayed complete immunity to fire. The two most famous of these "human salamanders"—in the middle ages the term *salamander* referred to a mythological lizard believed to live in fire—were Marie Sonnet and Gabrielle Moler. On one occasion, and in the presence of numerous witnesses, including Montgeron, Sonnet stretched herself on two chairs over a blazing fire and remained there for half an hour. Neither she nor her clothing showed any ill effects. In another instance she sat with her feet in a brazier full of burning coals. As with Brigham, her shoes and stockings burned off, but her feet were unharmed.[22]

Gabrielle Moler's exploits were even more dumbfounding. In addition to being impervious to the thrusts of swords and blows delivered by a shovel, she could stick her head into a roaring hearth fire and hold it there without suffering any injury. Eyewitnesses report that afterward her clothing was so hot it could barely be touched, yet her hair, eyelashes, and eyebrows were never so much as singed.[23] No doubt she was great fun at parties.

Actually the Jansenists were not the first convulsionary movement in France. In the late 1600s, when King Louis XIV tried to purge the country of the unabashedly Protestant Huguenots, a group of Huguenot resisters in the valley of the Cévennes and known as the Camisards displayed similar abilities. In an official report sent to Rome, one of the persecutors, a prior named Abbé du Chayla, complained that no matter what he did, he could not succeed in harming the Camisards. When he ordered them shot, the musket balls would be found flattened between their clothing and their skin. When he closed their hands upon burning coals, they were not harmed, and when he wrapped them head to toe in cotton soaked with oil and set them on fire, they did not burn.[24]

As if this weren't enough, Claris, the Camisard leader, ordered that a pyre be built and then climbed to the top of it to deliver an ecstatic speech. In the presence of six hundred witnesses he ordered the pyre be set on fire and continued to rant as the flames rose above his head. After the pyre was completely consumed, Claris remained, unharmed and with no mark of the fire on his hair or clothing. The head of the French troops sent to subdue the Camisards, a colonel named Jean Cavalier, was later exiled to England where he wrote a book on the event in 1707 entitled *A Cry from the Desert.*[25] As for Abbé du Chayla, he was eventually murdered by the Camisards during a retaliatory raid. Unlike some of them, he possessed no special invulnerability.[26]

Literally hundreds of credible accounts of fire immunity exist. It is reported that when Bernadette of Lourdes was in ecstasy she was also impervious to fire. According to witnesses, on one occasion her hand dropped so close to a burning candle while she was in trance that the flames licked around her fingers. One of the individuals present was Dr. Dozous, the municipal physician of Lourdes. Being of quick mind,

Dozous timed the event and noted that it was a full ten minutes before she came out of trance and removed her hand. He later wrote, "I saw it with my own eyes. But I swear, if anyone had tried to make me believe such a story I would have laughed him to scorn."[27]

On September 7, 1871, the *New York Herald* reported that Nathan Coker, an elderly Negro blacksmith living in Easton, Maryland, could handle red-hot metal without being harmed. In the presence of a committee that included several doctors, he heated an iron shovel until it was incandescent and then held it against the soles of his feet until it was cool. He also licked the edge of the red-hot shovel and poured melted lead shot in his mouth, allowing it to run over his teeth and gums until it solidified. After each of these feats the doctors examined him and found no trace of injury.[28]

While on a hunting trip in 1927 in the Tennessee mountains, K. R. Wissen, a New York physician, encountered a twelve-year-old boy who was similarly impervious. Wissen watched the boy handle red-hot irons out of a fireplace with impunity. The boy told Wissen he had discovered his ability by accident when he picked up a red-hot horseshoe in his uncle's blacksmith shop.[29] The pit of flaming embers the Grosvenors watched Mohotty walk through was twenty-feet long and measured 1328 degrees Fahrenheit on the *National Geographic* team's thermometers. In the May 1959 issue of the *Atlantic Monthly*, Dr. Leonard Feinberg of the University of Illinois reports witnessing another Ceylonese fire-walking ritual during which the natives carried red-hot iron pots on their heads without being harmed. In an article in *Psychiatric Quarterly*, psychiatrist Berthold Schwarz reports watching Appalachian Pentecostals hold their hands in an acetylene flame without being harmed,[30] and so on, and so on.

The Laws of Physics as Habits and Realities Both Potential and Real

Just as it is hard to imagine where the deflected energy goes in some of the examples of PK we have looked at, it is equally difficult to understand where the energy of a red-hot iron pot goes while the pot is resting flat against the hair and flesh of a Ceylonese native's head.

But if consciousness can mediate directly in the implicate order, it becomes a more tractable problem. Again, rather than being due to some undiscovered energy or law of physics (such as some kind of insulating force field) that operates *within* the framework of reality, it would result from activity on an even more fundamental level and involve the processes that create both the physical universe and the laws of physics in the first place.

Looked at another way, the ability of consciousness to shift from one entire reality to another suggests that the usually inviolate rule that *fire burns human flesh* may only be one program in the cosmic computer, but a program that has been repeated so often it has become one of nature's habits. As has been mentioned, according to the holographic idea, matter is also a kind of habit and is constantly born anew out of the implicate, just as the shape of a fountain is created anew out of the constant flow of water that gives it form. Peat humorously refers to the repetitious nature of this process as one of the universe's neuroses. "When you have a neurosis you tend to repeat the same pattern in your life, or do the same action, as if there's a memory built up and the thing is stuck with that," he says. "I tend to think things like chairs and tables are like that also. They're a sort of material neurosis, a repetition. But there is something subtler going on, a constant enfolding and unfolding. In this sense chairs and tables are just habits in this flux, but the flux is the reality, even if we tend only to see the habit."[31]

Indeed, given that the universe and the laws of physics that govern it are also products of this flux, then they, too, must be viewed as habits. Clearly they are habits that are deeply ingrained in the holomovement, but supernormal talents such as immunity to fire indicate that, despite their seeming constancy, at least some of the rules that govern reality can be suspended. This means the laws of physics are not set in stone, but are more like Shainberg's vortices, whirlpools of such vast inertial power that they are as fixed in the holomovement as our own habits and deeply held convictions are fixed in our thoughts.

Grof's proposal that altered states of consciousness may be required in order to make such changes in the implicate is also attested to by the frequency with which fire immunity is associated with heightened faith and religious zeal. The pattern that began to take shape in the last chapter continues, and its message becomes increasingly clear—the deeper and more emotionally charged our beliefs, the greater the changes we can make in both our bodies and reality itself.

At this point we might ask, if consciousness can make such extraordinary alterations under special circumstances, what role does it play in the creation of our day-to-day reality? Opinions are extremely varied. In private conversation Bohm admits to believing that the universe is all "thought" and reality exists only in what we think,[32] but again he prefers not to speculate about miraculous occurrences. Pribram is similarly reticent to comment on specific events but does believe a number of different potential realities exist and consciousness has a certain amount of latitude in choosing which one manifests. "I don't believe anything goes," he says, "but there are a lot of worlds out there that we don't understand."[33]

After years of firsthand experiences with the miraculous, Watson is bolder. "I have no doubt that reality is in a very large part a construct of the imagination. I am not speaking as a particle physicist or even as someone who is totally aware of what's going on in the frontier of that discipline, but I think we have the capacity to change the world around us in quite fundamental ways" (Watson, who was once enthusiastic about the holographic idea, is no longer convinced that *any* current theory in physics can adequately explain the supernormal abilities of the mind).[34]

Gordon Globus, a professor of psychiatry and philosophy at the University of California at Irvine, has a different but similar view. Globus thinks the holographic theory is correct in its assertion that the mind constructs concrete reality out of the raw material of the implicate. However, he has also been greatly influenced by anthropologist Carlos Castaneda's now famous otherworldly experiences with the Yaqui Indian shaman, Don Juan. In stark contrast to Pribram, he believes that the seemingly inexhaustible array of "separate realities" Castaneda experienced under Don Juan's tutelage—and indeed even the equally vast array of realities we experience during ordinary dreaming—indicate that there are an infinite number of potential realities enfolded in the implicate. Moreover, because the holographic mechanisms the brain uses to construct everyday reality are the same ones it uses to construct our dreams and the realities we experience during Castanedaesque altered states of consciousness, he believes all three types of reality are fundamentally the same.[35]

Does Consciousness Create Subatomic Particles or Not Create Subatomic Particles, That Is the Question

This difference of opinion indicates once again that the holographic theory is still very much an idea in the making, not unlike a newly formed Pacific island whose volcanic activity keeps it from having clearly defined shores. Although some might use this lack of consensus to criticize it, it should be remembered that Darwin's theory of evolution, certainly one of the most potent and successful ideas science has ever produced, is also still very much in a state of flux, and evolutionary theorists continue to debate its scope, interpretation, regulatory mechanisms, and ramifications.

The difference of opinion also reveals just how complex a puzzle miracles are. Jahn and Dunne offer yet another opinion on the role consciousness plays in the creation of day-to-day reality, and although it differs from one of Bohm's basic premises, because of the possible insight it offers into the process by which miracles are effected, it deserves our attention.

Unlike Bohm, Jahn and Dunne believe subatomic particles do *not* possess a distinct reality until consciousness enters the picture. "I think we have long since passed the place in high energy physics where we're examining the structure of a passive universe," Jahn states. "I think we're into the domain where the interplay of consciousness in the environment is taking place on such a primary scale that we are indeed creating reality by any reasonable definition of the term."[36]

As has been mentioned, this is the view held by most physicists. However, Jahn and Dunne's position differs from the mainstream in an important way. Most physicists would reject the idea that the interplay between consciousness and the subatomic world could in any way be used to explain PK, let alone miracles. In fact, the majority of physicists not only ignore any implications this interplay might have but actually behave as if it doesn't exist. "Most physicists develop a somewhat schizophrenic view," says quantum theorist Fritz Rohrlich of Syracuse University. "On the one hand they accept the standard interpretation of quantum theory. On the other they insist on the reality of quantum systems even when these are not observed."[37]

This bizarre I'm-not-going-to-think-about-it-even-when-I-know-it's-true attitude keeps many physicists from considering even the philo-sophical implications of quantum physics' most incredible findings. As N. David Mermin, a physicist at Cornell University, points out, physi-cists fall into three categories: a small minority is troubled by the philosophical implications; a second group has elaborate reasons why they are not troubled, but their explanations tend "to miss the point entirely"; and a third group has no elaborate explanations but also refuses to say why they aren't troubled. "Their position is unassail-able," says Mermin.[38]

Jahn and Dunne are not so timid. They believe that instead of discov-ering particles, physicists may actually be *creating* them. As evi-dence, they cite a recently discovered subatomic particle called an *anomalon*, whose properties vary from laboratory to laboratory. Imagine owning a car that had a different color and different features depending on who drove it! This is very curious and seems to suggest that an anomalon's reality depends on who finds/creates it.[39]

Similar evidence may also be found in another subatomic particle. In the 1930s Pauli proposed the existence of a massless particle called a *neutrino* to solve an outstanding problem concerning radioactivity. For years the neutrino was only an idea, but then in 1957 physicists discovered evidence of its existence. In more recent years, however, physicists have realized that if the neutrino possessed some mass, it would solve several even thornier problems than the one facing Pauli, and lo and behold in 1980 evidence started to come in that the neutrino had a small but measurable mass! This is not all. As it turned out, only laboratories in the Soviet Union discovered neutrinos with mass. Labo-ratories in the United States did not. This remained true for the better part of the 1980s, and although other laboratories have now duplicated the Soviet findings, the situation is still unresolved.[40]

Is it possible that the different properties displayed by neutrinos are due at least in part to the changing expectations and different cultural biases of the physicists who searched for them? If so, such a state of affairs raises an interesting question. If physicists do not discover the subatomic world but create it, why do some particles, such as elec-trons, appear to have a stable reality no matter who observes them? In other words, why does a physics student with no knowledge of an electron still discover the same characteristics that a seasoned physi-cist discovers?

One possible answer is that our perceptions of the world may not be based solely on the information we receive through our five senses.

As fantastic as this may sound, a very good case can be made for such a notion. Before explaining, I would like to relate an occurrence I witnessed in the middle 1970s. My father had hired a professional hypnotist to entertain a group of friends at his house and had invited me to attend the event. After quickly determining the hypnotic susceptibility of the various individuals present, the hypnotist chose a friend of my father's named Tom as his subject. This was the first time Tom had ever met the hypnotist.

Tom proved to be a very good subject, and within seconds the hypnotist had him in a deep trance. He then proceeded with the usual tricks performed by stage hypnotists. He convinced Tom there was a giraffe in the room and had Tom gaping in wonder. He told Tom that a potato was really an apple and had Tom eat it with gusto. But the highlight of the evening was when he told Tom that when he came out of trance, his teenage daughter, Laura, would be completely invisible to him. Then, after having Laura stand directly in front of the chair in which Tom was sitting, the hypnotist awakened him and asked him if he could see her.

Tom looked around the room and his gaze appeared to pass right through his giggling daughter. "No," he replied. The hypnotist asked Tom if he was certain, and again, despite Laura's rising giggles, he answered no. Then the hypnotist went behind Laura so he was hidden from Tom's view and pulled an object out of his pocket. He kept the object carefully concealed so that no one in the room could see it, and pressed it against the small of Laura's back. He asked Tom to identify the object. Tom leaned forward as if staring directly through Laura's stomach and said that it was a watch. The hypnotist nodded and asked if Tom could read the watch's inscription. Tom squinted as if struggling to make out the writing and recited both the name of the watch's owner (which happened to be a person unknown to any of us in the room) and the message. The hypnotist then revealed that the object was indeed a watch and passed it around the room so that everyone could see that Tom had read its inscription correctly.

When I talked to Tom afterward, he said that his daughter had been absolutely invisible to him. All he had seen was the hypnotist standing and holding a watch cupped in the palm of his hand. Had the hypnotist let him leave without telling him what was going on, he never would have known he wasn't perceiving normal consensus reality.

Obviously Tom's perception of the watch was not based on information he was receiving through his five senses. Where was he getting the information from? One explanation is that he was obtaining it telepathically from someone else's mind, in this case, the hypnotist's. The ability of hypnotized individuals to "tap" into the senses of other people has been reported by other investigators. The British physicist Sir William Barrett found evidence of the phenomenon in a series of experiments with a young girl. After hypnotizing the girl he told her that she would taste everything he tasted. "Standing behind the girl, whose eyes I had securely bandaged, I took up some salt and put it in my mouth; instantly she sputtered and exclaimed, 'What for are you putting salt in my mouth?' Then I tried sugar; she said 'That's better'; asked what it was like, she said 'Sweet.' Then mustard, pepper, ginger, et cetera were tried; each was named and apparently tasted by the girl when I put them in my own mouth."[41]

In his book *Experiments in Distant Influence* the Soviet physiologist Leonid Vasiliev cites a German study conducted in the 1950s that produced similar findings. In that study, the hypnotized subject not only tasted what the hypnotist tasted, but blinked when a light was flashed in the hypnotist's eyes, sneezed when the hypnotist took a whiff of ammonia, heard the ticking of a watch held to the hypnotist's ear, and experienced pain when the hypnotist pricked himself with a needle—all done in a manner that safeguarded against her obtaining the information through normal sensory cues.[42]

Our ability to tap into the senses of others is not limited to hypnotic states. In a now famous series of experiments physicists Harold Puthoff and Russell Targ of the Stanford Research Institute in California found that just about everyone they tested had a capacity they call "remote viewing," the ability to describe accurately what a distant test subject is seeing. They found that individual after individual could remote-view simply by relaxing and describing whatever images came into their minds.[43] Puthoff and Targ's findings have been duplicated by dozens of laboratories around the world, indicating that remote viewing is probably a widespread latent ability in all of us.

The Princeton Anomalies Research lab has also corroborated Puthoff and Targ's findings. In one study Jahn himself served as the receiver and tried to perceive what a colleague was observing in Paris, a city Jahn has never visited. In addition to seeing a bustling street, an image of a knight in armor came into Jahn's mind. It later turned out that the sender was standing in front of a government building ornamented with statuary of historical military figures, one of whom was a knight in armor.[44]

So it appears that we are deeply interconnected with each other in yet another way, a situation that is not so strange in a holographic universe. Moreover, these interconnections manifest even when we are not consciously aware of them. Studies have shown that when a person in one room is given an electric shock, it will register in the polygraph readings of a person in another room.[45] A light flashed in a test subject's eyes will register in the EEG readings of a test subject isolated in another room,[46] and even the blood volume of a test subject's finger changes—as measured by a plethysmograph, a sensitive indicator of autonomic nervous system functioning—when a "sender" in another room encounters the name of someone they know while reading a list composed mainly of names unknown to them.[47]

Given both our deep interconnectedness and our ability to construct entirely convincing realities out of information received via this interconnectedness, such as Tom did, what would happen if two or more hypnotized individuals tried to construct the same imaginary reality? Intriguingly, this question has been answered in an experiment conducted by Charles Tart, a professor of psychology at the Davis campus of the University of California. Tart found two graduate students, Anne and Bill, who could go into deep trance and were also skilled hypnotists in their own right. He had Anne hypnotize Bill and after he was hypnotized, he had Bill hypnotize her in return. Tart's reasoning was that the already powerful rapport that exists between hypnotist and subject would be strengthened by using this unusual procedure.

He was right. When they opened their eyes in this mutually hypnotized state everything looked gray. However, the grayness quickly gave way to vivid colors and glowing lights, and in a few moments they found themselves on a beach of unearthly beauty. The sand sparkled like diamonds, the sea was filled with enormous frothing bubbles and glistened like champagne, and the shoreline was dotted with translucent crystalline rocks pulsing with internal light. Although Tart could not see what Anne and Bill were seeing, from the way they were talking he quickly realized *they were experiencing the same hallucinated reality.*

Of course, this was immediately obvious to Anne and Bill and they set about to explore their newfound world, swimming in the ocean and studying the glowing crystalline rocks. Unfortunately for Tart they also stopped talking, or at least they stopped talking from Tart's perspective. When he questioned them about their silence they told him that in their shared dreamworld they *were* talking, a phenomenon

Tart feels involved some kind of paranormal communication between the two.

In session after session Anne and Bill continued to construct various realities, and all were as real, available to the five senses, and dimensionally realized, as anything they experienced in their normal waking state. In fact, Tart resolved that the worlds Anne and Bill visited were actually *more* real than the pale, lunar version of reality with which most of us must be content. As he states, after "they had been talking about their experiences to each other for some time, and found they had been discussing details of the experiences they had shared for which there were no verbal stimuli on the tapes, they felt they must have actually been 'in' the nonworldly locales they had experienced."[48]

Anne and Bill's ocean world is the perfect example of a holographic reality—a three-dimensional construct created out of interconnectedness, sustained by the flow of consciousness, and ultimately as plastic as the thought processes that engendered it. This plasticity was evident in several of its features. Although it was three-dimensional, its space was more flexible than the space of everyday reality and sometimes took on an elasticity Anne and Bill had no words to describe. Even stranger, although they were clearly highly skilled at sculpting a shared world outside themselves, they frequently forgot to sculpt their own bodies, and existed more often than not as floating faces or heads. As Anne reports, on one occasion when Bill told her to give him her hand, "I had to kind of conjure up a hand."[49]

How did this experiment in mutual hypnosis end? Sadly, the idea that these spectacular visions were somehow real, perhaps even more real than everyday reality, so frightened both Anne and Bill that they became increasingly nervous about what they were doing. They eventually stopped their explorations, and one of them, Bill, even gave up hypnosis entirely.

The extrasensory interconnectedness that allowed Anne and Bill to construct their shared reality might almost be viewed as a kind of field effect between them, a "reality-field" if you will. One wonders what would have happened if the hypnotist at my father's house had put all of us into a trance? In light of the evidence above, there is every reason to believe that if our rapport were deep enough, Laura would have become invisible to us all. We would have collectively constructed a reality-field of a watch, read its inscription, and been completely convinced that what we were perceiving was real.

If consciousness plays a role in the creation of subatomic particles, is it possible that our observations of the subatomic world are also reality-fields of a kind? If Jahn can perceive a suit of armor through the senses of a friend in Paris, is it any more farfetched to believe that physicists all around the world are unconsciously interconnecting with one another and using a form of mutual hypnosis similar to that used by Tart's subjects to create the consensus characteristics they observe in an electron? This possibility may be supported by another unusual feature of hypnosis. Unlike other altered states of consciousness, hypnosis is not associated with any unusual EEG patterns. Physiologically speaking, the mental state hypnosis most closely resembles is our normal waking consciousness. Does this mean that normal waking consciousness is itself a kind of hypnosis, and we are all constantly tapping into reality-fields?

Nobelist Josephson has suggested that something like this may be going on. Like Globus, he takes Castaneda's work seriously and has attempted to relate it to quantum physics. He proposes that objective reality is produced out of the collective memories of the human race while anomalous events, such as those experienced by Castaneda, are the manifestation of the individual will.[50]

Human consciousness may not be the only thing that participates in the creation of reality-fields. Remote viewing experiments have shown that people can accurately describe distant locations even when there are no human observers present at the locations.[51] Similarly, subjects can identify the contents of a seaied box randomly selected from a group of sealed boxes and whose contents are therefore completely unknown.[52] This means that we can do more than just tap into the senses of other people. We can also tap into reality itself to gain information. As bizarre as this sounds, it is not so strange when one remembers that in a holographic universe, consciousness pervades all matter, and "meaning" has an active presence in both the mental and physical worlds.

Bohm believes the ubiquitousness of meaning offers a possible explanation for both telepathy and remote viewing. He thinks both may actually be just different forms of psychokinesis. Just as PK is a resonance of meaning conveyed from a mind to an object, telepathy can be viewed as a resonance of meaning conveyed from a mind to a mind, says Bohm. In like manner, remote viewing can be looked at as a resonance of meaning conveyed from an object to a mind. "When harmony or resonance of 'meanings' is established, the action works both ways, so that the 'meanings' of the distant system could act in the viewer to produce a kind of inverse psychokinesis that would, in effect, transmit an image of that system to him," he states.[53]

Jahn and Dunne have a similar view. Although they believe reality is established only in the interaction of a consciousness with its environment, they are very liberal in how they define consciousness. As they see it, anything capable of generating, receiving, or utilizing information can qualify. Thus, animals, viruses, DNA, machines (artificially intelligent and otherwise), and so-called nonliving objects may all have the prerequisite properties to take part in the creation of reality.[54]

If such assertions are true, and we can obtain information not only from the minds of other human beings but from the living hologram of reality itself, psychometry—the ability to obtain information about an object's history simply by touching it—would also be explained. Rather than being inanimate, such an object would be suffused with its own kind of consciousness. Instead of being a "thing" that exists separately from the universe, it would be part of the interconnectedness of all things—connected to the thoughts of every person who ever came in contact with it, connected to the consciousness that pervades every animal and object that was ever associated with its existence, connected via the implicate to its own past, and connected to the mind of the psychometrist holding it.

You *Can* Get Something for Nothing

Do physicists play a role in the creation of subatomic particles? At present the puzzle remains unresolved, but our ability to interconnect with one another and conjure up realities that are as real as our normal waking reality is not the only clue that this may be the case. Indeed, the evidence of the miraculous indicates that we have scarcely even begun to fathom our talents in this area. Consider the following miraculous healing reported by Gardner. In 1982 an English physician named Ruth Coggin, working in Pakistan, was visited by a thirty-five-year-old Pakistani woman named Kamro. Kamro was eight months pregnant and for the better part of her pregnancy had suffered from bleeding and intermittent abdominal pain. Coggin recommended that she go into the hospital immediately, but Kamro refused.

Nonetheless, two days later her bleeding became so severe that she was admitted on an emergency basis.

Coggin's examination revealed that Kamro's blood loss had been "very heavy," and her feet and abdomen were pathologically swollen. The next day Kamro had "another heavy bleed," forcing Coggin to perform a cesarean section. As soon as Coggin opened the uterus even more copious amounts of dark blood flooded out and continued to flow so heavily it became clear that Kamro had virtually no clotting ability. By the time Coggin delivered Kamro's healthy baby daughter, "deep pools of unclotted blood" filled her bed and continued to flow from her incision. Coggin managed to obtain two pints of blood to transfuse the gravely anemic woman, but it was not nearly enough to replace the staggering loss. Having no other options, Coggin resorted to prayer.

She writes, "We prayed with the patient after explaining to her about Jesus in whose name we had prayed for her before the operation, and who was a great healer. I also told her that we were not going to worry. I had seen Jesus heal this condition before and was sure He was going to heal her."[55]

Then they waited.

For the next several hours Kamro continued to bleed, but instead of getting worse, her general condition stabilized. That evening Coggin prayed with Kamro again, and although her "brisk bleeding" continued unabated, she seemed unaffected by the loss. Forty-eight hours after the operation her blood finally began to clot and her recovery started in full. Ten days later she went home with her baby.

Although Coggin had no way of measuring Kamro's actual blood loss, she had no doubts that the young mother had lost more than her total blood volume during the surgery and the profuse bleeding that ensued. After Gardner examined the documentation of the case, he agreed. The trouble with this conclusion is that human beings cannot produce new blood fast enough to cover such catastrophic losses; if they could, many fewer people would bleed to death. This leaves one with the unsettling conclusion that Kamro's new blood must have materialized out of thin air.

The ability to create an infinitesimal particle or two pales in comparison to the materialization of the ten to twelve pints of blood necessary to replenish the average human body. And blood is not the only thing we can create out of thin air. In June of 1974, while traveling in Timor Timur, a small island in easternmost Indonesia, Watson encountered an equally confounding example of materialization. Although his original intention had been to visit a famous *matan do'ok*, a type of Indonesian wonder-worker who was said to be able to make it rain on demand, he was diverted by accounts of an unusually active *buan*, an evil spirit, wreaking havoc in a house in a nearby village.

The family living in the house consisted of a married couple, their two small boys, and the husband's unmarried younger half-sister. The couple and their children were typically Indonesian in appearance, with dark complexions and curly hair, but the half-sister, whose name was Alin, was physically very different and had a much lighter complexion and features that were almost Chinese, which accounted for her inability to obtain a husband. She was also treated with indifference by the family, and it was immediately plain to Watson that she was the source of the psychic disturbance.

That evening during dinner in the family's grass-roofed home, Watson witnessed several startling phenomena. First, without warning, the couple's eight-year-old boy screamed and dropped his cup on the table as the back of his hand began to bleed inexplicably. Watson, who was sitting next to the boy, examined his hand and saw that there was a semicircle of fresh punctures on it, like a human bite, but with a diameter larger than the boy's. Alin, always the odd person out, was busy at the fire opposite the boy when this occurred.

As Watson was examining the wounds, the lamp flame turned blue and abruptly flared up, and in the suddenly brighter light a shower of salt began to pour down over the food until it was completely covered and inedible. "It wasn't a sudden deluge, but a slow and deliberate action which lasted long enough for me to look up and see that it seemed to begin in midair, just about eye level, perhaps four feet over the table," says Watson.

Watson immediately leapt up from the table, but the show wasn't over. Suddenly a series of loud rapping sounds issued from the table, and it began to wobble. The family also jumped up and all watched as the table bucked "like the lid on a box containing some wild animal," and finally flipped over on its side. Watson first reacted by running out of the house with the rest of the family, but when he recovered his senses he returned and searched the room for evidence of any trickery that might account for the occurrence. He found none.[56]

The events that took place in the little Indonesian hut are classic examples of a poltergeist haunting, a type of haunting typified by mysterious sounds and psychokinetic activity rather than the appearances of ghosts or apparitions. Because poltergeists tend to center more around people, in this case Alin, rather than places, many parapsychologists believe they are actually manifestations of the unconscious psychokinetic ability of the person around whom they are most active. Even materialization has a long and illustrious history in the annals of poltergeist research. For instance, in his classic work on the subject, *Can We Explain the Poltergeist,* A. R. G. Owen, a fellow and lecturer in Mathematics at Trinity College, Cambridge, gives numerous examples of objects materializing out of thin air in poltergeist cases dating from A.D. 530 to modern times.[57] Small stones and not salt, however, are the objects that materialize most often.

In the Introduction I mentioned that I had experienced firsthand many of the paranormal phenomena that would be discussed in this book and would relate a few of my own experiences. It is thus time to come clean and confess that I know how Watson must have felt after witnessing the sudden onslaught of psychokinetic activity in the little Indonesian hut because when I was a child, the house in which my family had recently moved (a new house that my parents themselves had built) became the site of an active poltergeist haunting. Since our poltergeist left my family's home and followed me when I went away to college, and since its activity very definitely seemed connected to my moods—its antics becoming more malicious when I was angry or my spirits were low, and more impish and whimsical when my mood was brighter—I have always accepted the idea that poltergeists are manifestations of the unconscious psychokinetic ability of the person around whom they are most active.

This connection to my emotions displayed itself frequently. If I was in a good mood, I might wake up to find all of my socks draped over the house plants. If I was in a darker frame of mind, the poltergeist might manifest by hurling a small object across the room or occasionally even by breaking something. Over the years both I and various family members and friends witnessed a wide range of psychokinetic activity. My mother tells me that even when I was a toddler pots and pans had already begun to jump inexplicably from the middle of the kitchen table to the floor. I have written about some of these experiences in my book *Beyond the Quantum.*

I do not make these disclosures lightly. I am aware of how alien such occurrences are to most people's experience and fully understand the skepticism with which they will be greeted in some quarters. Nonetheless, I am compelled to talk about them because I think it is vitally important that we try to understand such phenomena and not just sweep them under the carpet.

Still it is with some trepidation that I admit that my own poltergeist also occasionally materialized objects. The materializations started when I was six years old, and inexplicable showers of gravel rained down on our roof at night. Later it took to pelting me *inside* my home with small polished stones and pieces of broken glass with edges worn like the shards of drift glass one finds on the beach. On rarer occasions it materialized other objects including coins, a necklace, and several odder trifles. Unfortunately, I usually did not see the actual materializations, but only witnessed their aftermath, such as when a pile of spaghetti noodles (sans sauce) fell on my chest one day while I was taking a nap in my New York apartment. Given that I was alone in a room with no open windows or doors, there was no one else in my apartment, and there was no sign that anyone had either cooked spaghetti or broken in to throw spaghetti at me, I can only assume that, for reasons unknown, the handful of cold spaghetti noodles that dropped out of midair and onto my chest materialized out of nowhere.

On a few occasions, however, I did see objects actually materialize. For example, in 1976 I was working in my study when I happened to look up and see a small brown object appear suddenly in midair just a few inches below the ceiling. As soon as it popped into existence it zoomed down at a sharp angle and landed at my feet. When I picked it up I saw that it was a piece of brown drift glass that originally might have been used in making beer bottles. It was not quite as spectacular as a shower of salt lasting several seconds, but it taught me that such things were possible.

Perhaps the most famous modern-day materializations are those produced by Sathya Sai Baba, a sixty-four-year-old Indian holy man living in a distant corner of the state of Andhra Pradesh in southern India. According to numerous eyewitnesses, Sai Baba is able to produce much more than salt and a few stones. He plucks lockets, rings, and jewelry out of the air and passes them out as gifts. He also materializes an endless supply of Indian delicacies and sweets, and out of his hands pour volumes of *vibuti,* or sacred ash. These events have been witnessed by literally thousands of individuals, including both scientists and magicians, and no one has ever detected any hint of trickery. One witness is psychologist Erlendur Haraldsson of the University of Iceland.

Haraldsson has spent over ten years studying Sai Baba and has published his findings in a recent book entitled *Modern Miracles: An Investigative Report on Psychic Phenomena Associated with Sathya Sai Baba.* Although Haraldsson admits that he cannot prove conclusively that Sai Baba's productions are not the result of deception and sleight of hand, he offers a large amount of evidence that strongly suggests something supernormal is taking place.

For starters, Sai Baba can materialize specific objects on request. Once when Haraldsson was having a conversation with him about spiritual and ethical issues, Sai Baba said that daily life and spiritual life should "grow together like a *double rudraksha.*" When Haraldsson asked what a double rudraksha was, neither Sai Baba nor the interpreter knew the English equivalent of the term. Sai Baba tried to continue with the discussion, but Haraldsson remained insistent. "Then suddenly, with a sign of impatience, Sai Baba closed his fist and waved his hand for a second or two. As he opened it, he turned to me and said: 'This is it.' In his palm was an acorn-like object. This was two rudrakshas grown together like a twin orange or a twin apple," says Haraldsson.

When Haraldsson indicated that he wanted to keep the double-seed as a memento, Sai Baba agreed, but first asked to see it again. "He enclosed the rudraksha in both his hands, blew on it, and opened his hands toward me. The double rudraksha was now covered, on the top and bottom, by two golden shields held together by a short golden chain. On the top was a golden cross with a small ruby affixed to it, and a tiny opening so that it could hang on a chain around the neck."[58] Haraldsson later discovered that double rudrakshas were extremely rare botanical anomalies. Several Indian botanists he consulted said they had never even seen one, and when he finally found a small, malformed specimen in a shop in Madras, the shopkeeper wanted the Indian equivalent of almost three hundred dollars for it. A London goldsmith confirmed that the gold in the ornamentation had a purity of at least twenty-two carats.

Such gifts are not rare. Sai Baba frequently hands out costly rings, jewels, and objects made of gold to the throngs who visit him daily and who venerate him as a saint. He also materializes vast quantities of food, and when the various delicacies he produces fall from his hands they are sizzling hot, so hot that people sometimes cannot even hold them. He can make sweet syrups and fragrant oils pour from his hands (and even his feet), and when he is finished there is no trace of the sticky substance on his skin. He can produce exotic objects such as grains of rice with tiny, perfectly carved pictures of Krishna on them, out-of-season fruits (a near impossibility in an area of the country that has no electricity or refrigeration), and anomalous fruits, such as apples that, when peeled, turn out to be an apple on one side and another fruit on the other.

Equally astonishing are his productions of sacred ash. Every time he walks among the crowds that visit him, prodigious amounts of it pour from his hands. He scatters it everywhere, into offered containers and outstretched hands, over heads, and in long serpentine trails on the ground. In a single transit of the grounds around his ashram he can produce enough of it to fill several drums. On one of his visits, Haraldsson, along with Dr. Karlis Osis, the director of research for the American Society for Psychical Research, actually saw some of the ash in the process of materializing. As Haraldsson reports, "His palm was open and turned downwards, and he waved his hand in a few quick, small circles. As he did, a grey substance appeared in the air just below his palm. Dr. Osis, who sat slightly closer, observed that this material first appeared entirely in the form of granules (that crumbled into ash when touched) and might have disintegrated earlier if Sai Baba had produced them by a sleight of hand that was undetectable to us."[59]

Haraldsson notes that Sai Baba's manifestations are not the result of mass hypnosis because he freely allows his open-air demonstrations to be filmed, and everything he does still shows up in the film. Similarly, the production of specific objects, the rarity of some of the objects, the hotness of the food, and the sheer volume of the materializations seem to rule against deception as a possibility. Haraldsson also points out that no one has ever come forth with any credible evidence that Sai Baba is faking his abilities. In addition, Sai Baba has been producing a continuous flow of objects for half a century, since he was fourteen, a fact that is further testament to both the volume of the materializations and the significance of his untarnished reputation. Is Sai Baba producing objects out of nothingness? At present the jury is still out, but Haraldsson makes it clear what his position is. He believes Sai Baba's demonstrations remind us of the "enormous potentials that may lie dormant somewhere within all human beings."[60]

Accounts of individuals who can materialize are not unknown in India. In his book *Autobiography of a Yogi*, Paramahansa Yogananda (1893–1952), the first eminent holy man of India to set up permanent residence in the West, describes his meetings with several Hindu ascetics who could materialize out-of-season fruits, gold plates, and other objects. Interestingly, Yogananda cautioned that such powers, or *siddis*, are not always evidence that the person possessing them is spiritually evolved. "The world [is] nothing but an objectivized dream," says Yogananda, and "whatever your powerful mind believes very intensely instantly comes to pass."[61] Have such individuals discovered a way to tap just a little of the enormous sea of cosmic energy that Bohm says fills every cubic centimeter of empty space?

A remarkable series of materializations that has received even greater confirmation than that bestowed by Haraldsson on Sai Baba was produced by Therese Neumann. In addition to her stigmata, Neumann also displayed *inedia*, the supernormal ability to live without food. Her inedia began in 1923 when she "transferred" the throat disease of a young priest to her own body and subsisted solely on liquids for several years. Then, in 1927, she gave up both food and water entirely.

When the local bishop in Regensburg first learned of Neumann's fast, he sent a commission into her home to investigate. From July 14, 1927, to July 29, 1927, and under the supervision of a medical doctor named Seidl, four Franciscan nursing sisters scrutinized her every move. They watched her day and night, and the water she used for washing and rinsing her mouth was carefully measured and weighed. The sisters discovered several unusual things about Neumann. She never went to the bathroom (even after a period of six weeks she only had one bowel movement, and the excrement, examined by a Dr. Reismanns, contained only a small amount of mucus and bile, but no traces of food). She also showed no signs of dehydration, even though the average human expels about four hundred grams (fourteen ounces) of water daily in the air he or she exhales, and a like amount through the pores. And her weight remained constant; although she lost nearly nine pounds (in blood) during the weekly opening of her stigmata, her weight returned to normal within a day or two later.

At the end of the inquiry Dr. Seidl and the sisters were completely convinced that Neumann had not eaten or drunk a thing for the entire fourteen days. The test seems conclusive, for while the human body can survive two weeks without food, it can rarely survive half that time without water. Yet this was nothing for Neumann; *she did not eat or drink a thing for the next thirty-five years.* So it appears that she was not only materializing the enormous amount of blood necessary to perpetuate her stigmata, but also regularly materializing the water and nutrients she needed to stay alive and in good health. Inedia is not unique to Neumann. In *The Physical Phenomena of Mysticism*, Thurston gives several examples of stigmatists who went for years without eating or drinking.

Materialization may be more common than we realize. Compelling accounts of bleeding statues, paintings, icons, and even rocks that have historical or religious significance abound in the literature on the miraculous. There are also dozens of stories of Madonnas and other icons shedding tears. A virtual epidemic of "weeping Madonnas" swept Italy in 1953.[62] And in India, followers of Sai Baba showed Haraldsson pictures of the ascetic that were miraculously exuding sacred ash.

Changing the Whole Picture

In a way materialization challenges our conventional ideas about reality most of all, for although we can, with effort, hammer things such as PK into our current world view, the creation of an object out of thin air rocks the very foundation of that world view. Still, it is not all the mind can do. So far we have looked at miracles that involve only "parts" of reality—examples of people psychokinetically moving parts around, of people altering parts (the laws of physics) to make themselves immune to fire, and of people materializing parts (blood, salt, stones, jewelry, ash, nutrients, and tears). But if reality is really an unbroken whole, why do miracles seem to involve only parts?

If miracles are examples of the mind's own latent abilities, the answer, of course, is because we ourselves are so deeply programmed to see the world in terms of parts. This implies that if we were not so inculcated in thinking in terms of parts, if we viewed the world differently, miracles would also be different. Rather than finding so many examples of miracles in which the parts of reality had been transformed, we would find more instances in which the whole of reality had been transformed. In fact a few such examples exist, but they are rare and offer an even graver challenge to our conventional ideas about reality than materializations do.

Watson provides one. While he was in Indonesia he also encountered another young woman with power. The woman's name was Tia, but unlike Alin's power, hers did not seem to be an expression of an unconscious psychic gift. Instead it was consciously controlled and stemmed from Tia's natural connection to forces that lie dormant in most of us. Tia was, in short, a shaman in the making. Watson witnessed many examples of her gifts. He saw her perform miraculous healings, and once, when she was engaged in a power struggle with the local Moslem religious leader, he saw her use the power of her mind to set the minaret of the local mosque on fire.

But he witnessed one of Tia's most awesome displays when he accidentally stumbled upon her talking with a little girl in a shady grove of *kenari* trees. Even at a distance, Watson could tell from Tia's gestures that she was trying to communicate something important to the child. Although he could not hear their conversation, he could tell from her air of frustration that she was not succeeding. Finally, she appeared to get an idea and started an eerie dance.

Entranced, Watson continued to watch as she gestured toward the trees, and although she scarcely seemed to move, there was something hypnotic about her subtle gesticulations. Then she did something that both shocked and dismayed Watson. She caused the entire grove of trees suddenly to blink out of existence. As Watson states, "One moment Tia danced in a grove of shady *kenari;* the next she was standing alone in the hard, bright light of the sun."[63]

A few seconds later she caused the grove to reappear, and from the way the little girl leapt to her feet and rushed around touching the trees, Watson was certain that she had shared the experience also. But Tia was not finished. She caused the grove to blink on and off several times as both she and the little girl linked hands, dancing and giggling at the wonder of it all. Watson simply walked away, his head reeling.

In 1975 when I was a senior at Michigan State University I had a similarly profound and reality-challenging experience. I was having dinner with one of my professors at a local restaurant, and we were discussing the philosophical implications of Carlos Castaneda's experiences. In particular our conversation centered around an incident Castaneda relates in *Journey to Ixtlan.* Don Juan and Castaneda are in the desert at night searching for a spirit when they come upon a creature that looks like a calf but has the ears of a wolf and the beak of a bird. It is curled up and screaming as if in the throes of an agonizing death.

At first Castaneda is terrified, but after telling himself that what he is seeing can't possibly be real, his vision changes and he sees that the dying spirit is actually a fallen tree branch trembling in the wind. Castaneda proudly points out the thing's true identity, but as usual the old Yaqui shaman rebukes him. He tells Castaneda that the branch *was* a dying spirit while it was alive with power, but that it had transformed into a tree branch when Castaneda doubted its existence. However, he stresses that both realities were equally real.

In my conversation with my professor, I admitted that I was intrigued by Don Juan's assertion that two mutually exclusive realities could each be real and felt that the notion could explain many paranormal events. Moments after discussing this incident we left the restaurant and, because it was a clear summer night, we decided to stroll. As we continued to converse I became aware of a small group of people walking ahead of us. They were speaking an unrecognizable foreign language, and from their boisterous behavior it appeared that they were drunk. In addition, one of the women was carrying a green umbrella, which was strange because the sky was totally cloudless and there had been no forecast of rain.

Not wanting to collide with the group, we dropped back a little, and as we did, the woman suddenly began swinging the umbrella in a wild and erratic manner. She traced out huge arcs in the air, and several times as she spun around, the tip of the umbrella nearly grazed us. We slowed our pace even more, but it became increasingly apparent that her performance was designed to attract our attention. Finally, after she had our gaze firmly fixed on what she was doing, she held the umbrella with both hands over her head and then threw it dramatically at our feet.

We both stared at it dumbly, wondering why she had done such a thing, when suddenly something remarkable began to happen. The umbrella did something that I can only describe as "flickering" like a lantern flame about to go out. It emitted an odd, crackling sound like the sound of cellophane being crumpled, and in a dazzling array of sparkling, multicolored light, its ends curled up, its color changed, and it reshaped itself into a gnarled, brown-gray stick. I was so stunned I didn't say anything for several seconds. My professor spoke first and said in a quiet, shocked voice that she had thought the object had been an umbrella. I asked her if she had seen something extraordinary happen and she nodded. We both wrote down what we thought had transpired and our accounts matched exactly. The only vague difference in our descriptions was that my professor said the umbrella had "sizzled" when it transformed into a stick, a sound not too terribly dissimilar from the crackly sound of cellophane being crumpled.

What Does It All Mean?

This incident raises many questions for which I have no answers. I do not know who the people were who threw the umbrella at our feet, or if they were even aware of the magical transformation that took place as they strolled away, although the woman's bizarre and seemingly purposeful performance suggests that they were not completely unwitting. Both my professor and I were so transfixed by the magical transformation of the umbrella that by the time we had the presence of mind to ask them, they were long gone. I do not know why the event happened, save that it seems obvious it was connected in some way to our talk about Castaneda encountering a similar occurrence.

I do not even know why I have had the privilege of experiencing so many paranormal occurrences, save that it appears to be related to the fact that I was born with a great deal of native psychic ability. As an adolescent I started having vivid and detailed dreams about events that would later happen. I often knew things about people I had no right knowing. When I was seventeen I spontaneously developed the ability to see an energy field, or "aura," around living things, and to this day can often determine things about a person's health by the pattern and colors of the mist of light that I see surrounding them. Above and beyond that, all I can say is that we are all gifted with different aptitudes and qualities. Some of us are natural artists. Some dancers. I seem to have been born with the chemistry necessary to trigger shifts in reality, to catalyze somehow the forces required to precipitate paranormal events. I am grateful for this capacity because it has taught me a great deal about the universe, but I do not know why I have it.

What I do know is that the "umbrella incident," as I have come to call it, entailed a radical alteration in the world. In this chapter we have looked at miracles that have involved increasingly greater shifts in reality. PK is easier for us to fathom than the ability to pluck an object out of the air, and the materialization of an object is easier for most of us to accept than the appearance and disappearance of an entire grove of trees, or the paranormal appearance of a group of people capable of transmogrifying matter from one form into another. More and more these incidents suggest that reality is, in a very real sense, a hologram, a construct.

The question becomes, Is it a hologram that is relatively stable for

long periods of time and subject to only minimal alterations by consciousness, as Bohm suggests? Or is it a hologram that only seems
stable, but under special circumstances can be changed and reshaped
in virtually limitless ways, as the evidence of the miraculous suggests? Some researchers who have embraced the holographic idea
believe the latter is the case. For example, Grof not only takes materialization and other extreme paranormal phenomena seriously, but
feels that reality is indeed cloud-built and pliant to the subtle authority
of consciousness. "The world is not necessarily as solid as we perceive
it," he says.[64]

Physicist William Tiller, head of the Department of Materials Science at Stanford University and another supporter of the holographic
idea, agrees. Tiller thinks reality is similar to the "holodeck" on the
television show *Star Trek: The Next Generation.* In the series, the
holodeck is an environment in which occupants can call up a holographic simulation of literally any reality they desire, a lush forest, a
bustling city. They can also change each simulation in any way they
want, such as cause a lamp to materialize or make an unwanted table
disappear. Tiller thinks the universe is also a kind of holodeck created
by the "integration" of all living things. "We've created it as a vehicle
of experience, and we've created the laws that govern it," he asserts.
"And when we get to the frontiers of our understanding, we can in
fact shift the laws so that we're also creating the physics as we go
along."[65]

If Tiller is right and the universe is an enormous holodeck, the
ability to materialize a gold ring or cause a grove of *kenari* trees to
flick on and off is no longer so strange. Even the umbrella incident can
be viewed as a temporary aberration in the holographic simulation we
call ordinary reality. Although my professor and I were unaware that
we possessed such an ability, it may be that the emotional fervor of
our discussion about Castaneda caused our unconscious minds to
change the hologram of reality to better reflect what we were believing at the moment. Given Ullman's assertion that our psyche is constantly trying to teach us things we are unaware of in our waking
state, our unconscious may even be programmed to produce occasionally such miracles in order to offer us glimpses of reality's true nature,
to show us that the world we create for ourselves is ultimately as
creatively infinite as the reality of our dreams.

Saying that reality is created by the integration of all living things
is really no different from saying that the universe is comprised of

reality fields. If this is true, it explains why the reality of some sub-atomic particles, such as electrons, seems relatively fixed, while the reality of others, such as anomalons, appears to be more plastic. It may be that the reality fields we now perceive as electrons became part of the cosmic hologram long ago, perhaps long before human beings were even part of the integration of all things. Hence, electrons may be so deeply ingrained in the hologram they are no longer as susceptible to the influence of human consciousness as other newer reality fields. Similarly, anomalons may vary from lab to lab *because* they are more recent reality fields and are still inchoate, still flounder-ing around in search of an identity, as it were. In a sense, they are like the champagne beach Tart's subjects perceived while it was still in its gray state and had not yet fully coalesced out of the implicate.

This may also explain why aspirin helps prevent heart attacks in Americans, but not in the British. It, too, may be a relatively recent reality field and one that is still in the making. There is even evidence that the ability to materialize blood is a comparatively recent reality field. Rogo notes that accounts of blood miracles began with the fourteenth-century miracle of San Gennaro. The fact that no blood miracles are known to predate San Gennaro seems to indicate that the ability flickered into existence at that time. Once it was thus estab-lished it would be easier for others to tap into the reality field of its possibility, which may explain why there have been numerous blood miracles since San Gennaro, but none before.

Indeed, if the universe is a holodeck, all things that appear stable and eternal, from the laws of physics to the substance of galaxies, would have to be viewed as reality fields, will-o'-the-wisps no more or less real than the props in a giant, mutually shared dream. All perma-nence would have to be looked at as illusory, and only consciousness would be eternal, the consciousness of the living universe.

Of course, there is one other possibility. It may be that only anoma-lous events, such as the umbrella incident, are reality fields, and the world at large is still every bit as stable and unaffected by conscious-ness as we have been taught to believe. The problem with this assump-tion is that it can never be proved. The only litmus test we have of determining whether something is real, say a purple elephant that has just strolled into our living room, is to find out if other people can see it as well. But once we admit that two or more people can create a reality—whether it is a transforming umbrella or a vanishing grove of *kenari* trees—we no longer have any way of proving that everything else in the world is not created by the mind. It all boils down to a matter of personal philosophy.

And personal philosophies vary. Jahn prefers to think that only the reality created by the interactions of consciousness are real. "The question of whether there's an 'out there' out there is abstract. If we have no way of verifying the abstraction, there is no profit in attempting to model it," he says.[66] Globus, who willingly admits that reality is a construct of consciousness, prefers to think that there is a world beyond the bubble of our perceptions. "I'm interested in nice theories," he says, "and a nice theory postulates existence."[67] However, he admits that this is merely his bias, and there is no empirical way to prove such an assumption.

As for me, as a result of my own experiences I agree with Don Juan when he states, "We are perceivers. We are an awareness; we are not objects; we have no solidity. We are boundless. The world of objects and solidity is a way of making our passage on earth convenient. It is only a description that was created to help us. We, or rather our *reason*, forget that the description is only a description and thus we entrap the totality of ourselves in a vicious circle from which we rarely emerge in our lifetime."[68]

Put another way, there is *no* reality above and beyond that created by the integration of all consciousnesses, and the holographic universe can potentially be sculpted in virtually limitless ways by the mind.

If this is true, the laws of physics and the substance of galaxies are not the only things that are reality fields. Even our bodies, the vehicles of our consciousness in this life, would have to be looked upon as no more or less real than anomalons and champagne beaches. Or as Keith Floyd, a psychologist at Virginia Intermont College and another supporter of the holographic idea, states, "Contrary to what everyone knows is so, it may not be the brain that produces consciousness, but rather consciousness that creates the appearance of the brain—matter, space, time and everything else we are pleased to interpret as the physical universe."[69]

This is perhaps most disturbing of all, for we are so deeply convinced that our bodies are solid and objectively real it is difficult for us even to entertain the idea that we, too, may be no more than will-o'-the-wisps. But there is compelling evidence that this is also the case. Another phenomenon often associated with saints is *bilocation*, or the ability to be in two places at once. According to Haraldsson, Sai Baba does biolocation one better. Numerous witnesses have reported watching him snap his fingers and vanish, instantly reappearing a hundred or more yards away. Such incidents very much suggest that our bodies are not objects, but holographic projections that can blink "off" in one location and "on" in another with the same ease that an image might vanish and reappear on a video screen.

An incident that further underscores the holographic and immaterial nature of the body can be found in phenomena produced by an Icelandic medium named Indridi Indridason. In 1905 several of Iceland's leading scientists decided to investigate the paranormal and chose Indridason as one of their subjects. At the time, Indridason was just a country bumpkin with no previous experience with things psychic, but he quickly proved to be a spectacularly talented medium. He could go into trance quickly and produce dramatic displays of PK. But most bizarre of all, sometimes while he was deep in trance, different parts of his body would completely dematerialize. As the astonished scientists watched, an arm or a hand would fade out of existence, only to rematerialize before he awakened.[70]

Such events again offer us a tantalizing glimpse of the enormous potentialities that may lie dormant in all of us. As we have seen, our current scientific understanding of the universe is completely incapable of explaining the various phenomena we have examined in this chapter and therefore has no choice but to ignore them. However, if researchers such as Grof and Tiller are correct and the mind is able to intercede in the implicate order, the holographic plate that gives birth to the hologram we call the universe, and thus create any reality or laws of physics that it wants to, then not only are such things possible, but virtually anything is possible.

If this is true, the apparent solidity of the world is only a small part of what is available to our perception. Although most of us are indeed entrapped in our current description of the universe, a few individuals do have the ability to see beyond the world's solidity. In the next chapter we will take a look at some of these individuals and examine what they see.

6

Seeing Holographically

We human beings consider ourselves to be made up of "solid matter." Actually, *the physical body is the end product,* so to speak, of the subtle information fields, which mold our physical body as well as all physical matter. These fields are holograms which change in time (and are) outside the reach of our normal senses. This is what clairvoyants perceive as colorful egg-shaped halos or auras surrounding our physical bodies.

—Itzhak Bentov
Stalking the Wild Pendulum

A number of years ago I was walking along with a friend when a street sign caught my attention. It was simply a No Parking sign and seemed no different from any of the other No Parking signs that dotted the city streets. But for some reason it held me transfixed. I wasn't even aware that I was staring at it until my friend suddenly exclaimed, "That sign is misspelled!" Her announcement snapped me out of my reverie, and as I watched, the *i* in the word *Parking* quickly changed into an *e*.

What happened was that my mind was so accustomed to seeing the sign spelled correctly that my unconscious edited out what was there and made me see what it expected to be there. My friend, as it turned out, had also seen the sign spelled correctly at first, which was why she had such a vocal reaction when she realized it was misspelled. We continued to walk on, but the incident bothered me. For the first time I realized that the eye/brain is not a faithful camera, but tinkers with the world before it gives it to us.

Neurophysiologists have long been aware of this fact. In his early studies of vision, Pribram discovered that the visual information a monkey receives via its optic nerves does not travel directly into its visual cortex, but is first filtered through other areas of its brain.[1] Numerous studies have shown that the same is true of human vision. Visual information entering our brains is edited and modified by our temporal lobes before it is passed on to our visual cortices. Some studies suggest that less than 50 percent of what we "see" is actually based on information entering our eyes. The remaining 50 percent plus is pieced together out of our expectations of what the world should look like (and perhaps out of other sources such as reality fields). The eyes may be visual organs, but it is the brain that sees.

This is why we don't always notice when a close friend shaves off his mustache, and why our house always looks strangely different when we return to it after a vacation. In both instances we are so used to responding to what we think is there, we don't always see what really is there.

Even more dramatic evidence of the role the mind plays in creating what we see is provided by the eye's so-called blind spot. In the middle of the retina, where the optic nerve connects to the eye, we have a blind spot where there are no photoreceptors. This can be quickly demonstrated with the illustration shown in figure 15.

Even when we look at the world around us we are totally unaware that there are gaping holes in our vision. It doesn't matter whether we are gazing at a blank piece of paper or an ornate Persian carpet. The brain artfully fills in the gaps like a skilled tailor reweaving a hole in a piece of fabric. What is all the more remarkable is that it reweaves the tapestry of our visual reality so masterfully we aren't even aware that it is doing so.

This leads to a disturbing question. If we are seeing less than half of what is out there, what is out there that we are not seeing? What misspelled street signs and blind spots are escaping our attention completely? Our technological prowess provides us with a few answers. For example, although spiderwebs look drab and white to us, we now know that to the ultraviolet-sensitive eyes of the insects for whom they were designed, they are actually brightly colored and hence alluring. Our technology also tells us that fluorescent lamps do

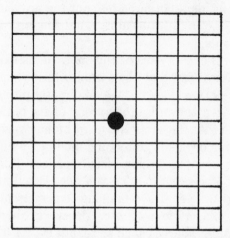

FIGURE 15. To demonstrate how our brains construct what we perceive as reality, hold the illustration at eye level, close your left eye, and stare at the circle in the middle of the grid with your right eye. Slowly move the book back and forth along the line of your vision until the star vanishes (about 10 to 15 inches). The star disappears because it is falling on your blind spot. Now close your right eye and stare at the star. Move the book back and forth until the circle in the middle of the grid vanishes. When it does, notice that although the circle disappears, all the lines of the grid remain intact. This is because your brain is filling in what it thinks should be there.

not continuously provide light, but are actually flickering on and off at a rate that is just a little too fast for us to discern. Yet this unsettling strobelike effect is quite visible to honeybees, who must be able to fly at breakneck speed over a meadow and still see every flower that whizzes by.

But are there other important aspects of reality that we are not seeing, aspects that are beyond even our technological grasp? According to the holographic model, the answer is yes. Remember that in Pribram's view, reality at large is really a frequency domain, and our brain is a kind of lens that converts these frequencies into the objective world of appearances. Although Pribram began by studying the frequencies of our normal sensory world, such as frequencies of sound and light, he now uses the term *frequency domain* to refer to the interference patterns that compose the implicate order.

Pribram believes there may be all kinds of things out there in the frequency domain that we are not seeing, things our brains have learned to edit out regularly of our visual reality. He thinks that when mystics have transcendental experiences, what they are really doing is catching glimpses of the frequency domain. "Mystical experience makes sense when one can provide the mathematical formulas that take one back and forth between the ordinary world, or 'image-object' domain, and the 'frequency' domain," he states.[2]

The Human Energy Field

One mystical phenomenon that appears to involve the ability to see reality's frequency aspects is the aura, or human energy field. The notion that there is a subtle field of energy around the human body, a halolike envelope of light that exists just beyond normal human perception, can be found in many ancient traditions. In India, sacred writings that date back over five thousand years refer to this life energy as *prana*. In China, since the third millennium B.C., it has been called *ch'i* and is believed to be the energy that flows through the acupuncture meridian system. Kabbalah, a Jewish mystical philosophy that arose in the sixth century B.C., calls this vital principle *nefish* and teaches that an egg-shaped bubble of iridescence surrounds every human body. In their book *Future Science*, writer John White and parapsychologist Stanley Krippner list 97 different cultures that refer to the aura with 97 different names.

Many cultures believe the aura of an extremely spiritual individual is so bright it is visible even to normal human perception, which is why so many traditions, including Christian, Chinese, Japanese, Tibetan, and Egyptian, depict saints as having halos or other circular symbols around their heads. In his book on miracles Thurston devotes an entire chapter to accounts of luminous phenomena associated with Catholic saints, and both Neumann and Sai Baba are reported to have occasionally had visible auras of light around them. The great Sufi mystic Hazrat Inayat Khan, who died in 1927, is said to have sometimes given off so much light that people could actually read by it.[3]

Under normal circumstances, however, the human energy field is visible only to individuals who have a specially developed capacity to see it. Sometimes people are born with the ability. Sometimes it develops spontaneously at a certain point in a person's life, as it did in my case, and sometimes it develops as the result of some practice or discipline, often of a spiritual nature. The first time I saw the distinctive mist of light around my arm I thought it was smoke and jerked my arm up to see if I had somehow caught my sleeve on fire. Of course, I hadn't and quickly discovered that the light surrounded my entire body and formed a nimbus around everyone else's as well.

According to some schools of thought the human energy field has a number of distinct layers. I do not see layers in the field and have no personal basis to judge if this is true or not. These layers are actually said to be three-dimensional energy bodies that occupy the same space as the physical body but are of increasingly larger size so that they only look like layers, or strata, as they extend outward from the body.

Many psychics assert that there are seven main layers, or subtle bodies, each progressively less dense than the one before it, and each increasingly more difficult to see. Different schools of thought refer to these energy bodies by different names. One common system of no- menclature refers to the first four as the etheric body; the astral, or emotional body; the mental body, and the causal, or intuitive body. It is generally believed that the etheric body, the body that is closest in size to the physical body, is a kind of energy blueprint and is involved in guiding and shaping the growth of the physical body. As their names suggest, the next three bodies are related to emotional, mental, and intuitive processes. Virtually no one agrees on what to call the remaining three bodies, although it is commonly agreed that they have to do with the soul and higher spiritual functioning.

According to Indian yogic literature, and to many psychics as well, we also have special energy centers in our body. These focal points of subtle energy are connected to endocrine glands and major nerve centers in the physical body, but also extend up and into the energy field. Because they resemble spinning vortices of energy when they are looked at head-on, yogic literature refers to them as *chakras*, from the Sanskrit word for "wheel," and this term is still used today.

The crown chakra, an important chakra that originates in the upper- most tip of the brain and is associated with spiritual awakening, is often described by clairvoyants as looking like a little cyclone whirling in the energy field on top of the head, and it is the only chakra I see clearly. (My own abilities appear to be too rudimentary to permit me to see the other chakras.) It ranges from a few inches to a foot or more in height. When people are in a joyous state, this whirlwind of energy grows taller and brighter, and when they dance, it bobs and sways like a candle flame. I've often wondered if this was what the apostle Luke was seeing when he described the "flame of the Pentecost," the tongues of fire that appeared on the heads of the apostles when the Holy Ghost descended on them.

The human energy field is not always bluish white, but can possess various colors. According to talented psychics, these colors, their muddiness or intensity, and their location in the aura are related to a person's mental state, emotional state, activity, health, and assorted other factors. I can only see colors occasionally and sometimes can interpret their meaning, but again my abilities in this area are not terribly advanced.

One person who does have advanced abilities is therapist and healer Barbara Brennan. Brennan began her career as an atmospherics physicist working for NASA at the Goddard Space Flight Center, and later left to become a counselor. Her first inkling that she was psychic came when she was a child and discovered she could walk blindfolded through the woods and avoid the trees simply by sensing their energy fields with her hands. Several years after she became a counselor, she began seeing halos of colored light around people's heads. After overcoming her initial shock and skepticism, she set about to develop the ability and eventually discovered she had an extraordinary natural talent as a healer.

Brennan not only sees the chakras, layers, and other fine structures of the human energy field with exceptional clarity, but can make startlingly accurate medical diagnoses based on what she sees. After looking at one woman's energy field, Brennan told her there was something abnormal about her uterus. The woman then told Brennan that her doctor had discovered the same problem, and it had already caused her to have one miscarriage. In fact, several physicians had recommended a hysterectomy and that was why she was seeking Brennan's counsel. Brennan told her that if she took a month off and took care of herself, her problem would clear up. Brennan's advice turned out to be correct, and a month later the woman's physician confirmed that her uterus had returned to normal. A year later the woman gave birth to a healthy baby boy.[4]

In another case Brennan was able to see that a man had problems performing sexually because he had broken his coccyx (tailbone) when he was twelve. The still out-of-place coccyx was applying undue pressure to his spinal column, and this in turn was causing his sexual dysfunction.[5]

There seems to be little Brennan cannot pick up by looking at the human energy field. She says that in its early stages cancer looks gray-blue in the aura, and as it progresses, it turns to black. Eventually, white spots appear in the black, and if the white spots sparkle and begin to look as if they are erupting from a volcano, it means the cancer has metastasized. Drugs such as alcohol, marijuana, and cocaine are also detrimental to the brilliant, healthy colors of the aura and create what Brennan calls "etheric mucus." In one instance she was able to tell a startled client which nostril he habitually used to snort cocaine because the field over that side of his face was always gray with the sticky etheric mucus.

Prescription drugs are not exempt, and often cause dark areas to form in the energy field over the liver. Potent drugs such as chemotherapy "clog" the entire field, and Brennan says she has even seen auric traces of the supposedly harmless radiopaque dye used to diagnose spinal injuries, a full ten years after it has been injected into a person's spine. According to Brennan, a person's psychological condition is also reflected in their energy field. An individual with psychopathic tendencies has a top-heavy aura. The energy field of a masochistic personality is coarse and dense and is more gray than blue. The field of a person with a rigid approach to life is also coarse and grayish, but with most of its energy concentrated on the outer edge of the aura, and so on.

Brennan says that illness can actually be caused by tears, blockages, and imbalances in the aura, and by manipulating these dysfunctional areas with her hands and her own energy field, she can greatly enhance a person's own healing processes. Her talents have not gone unnoticed. Swiss psychiatrist and thanatologist Elisabeth Kubler-Ross says Brennan is "probably one of the best spiritual healers in the Western Hemisphere."[6] Bernie Siegel is equally laudatory: "Barbara Brennan's work is mind opening. Her concepts of the role disease plays and how healing is achieved certainly fit in with my experience."[7]

As a physicist, Brennan is keenly interested in describing the human energy field in scientific terms and believes Pribram's assertion that there is a frequency domain beyond our field of normal perception is the best scientific model we have so far for understanding the phenomenon. "From the point of view of the holographic universe, these events [the aura and the healing forces required to manipulate its energies] emerge from frequencies that transcend time and space; they don't have to be transmitted. They are potentially simultaneous and everywhere," she says.[8]

That the human energy field exists everywhere and is nonlocal until it is plucked out of the frequency domain by human perception is evidenced in Brennan's discovery that she can read a person's aura even when the person is many miles distant. The longest-distance aura reading she has done so far was during a telephone conversation between New York City and Italy. She discusses this, as well as many other aspects of her remarkable abilities, in her recent and fascinating book *Hands of Light.*

The Energy Field of the Human Psyche

Another gifted psychic who can see the aura in great detail is Los Angeles–based "human energy field consultant" Carol Dryer. Dryer says she has been able to see auras for as long as she can remember, and indeed it was quite some time before she realized other people couldn't see auras. Her ignorance in this regard frequently landed her in trouble as a child when she would tell her parents intimate details about their friends, things she had no apparent way of knowing.

Dryer makes her living as a psychic, and in the past decade and a half has seen over five thousand clients. She is well known in the media because her client list includes many celebrities such as Tina Turner, Madonna, Rosanna Arquette, Judy Collins, Valerie Harper, and Linda Gray. But even the star power of her client list does not begin to convey the true extent of her talent. For instance, Dryer's client list also includes physicists, noted journalists, archaeologists, lawyers, and politicians, and she has used her abilities to assist the police and frequently does consultation work for psychologists, psychiatrists, and medical doctors.

Like Brennan, Dryer can give long-distance readings, but prefers to be in the same room with the person. She can also see a person's energy field as well with her eyes closed as she can with her eyes open. In fact, she generally keeps her eyes closed during a reading to help her concentrate solely on the energy field. This does not mean that she sees the aura only in her mind's eye. "It's always in front of me as if I'm looking at a movie or a play," says Dryer. "It's as real as the room I'm sitting in. Actually, it's more real and more brightly colored."[9]

However, she does not see the precise stratified layers described by other clairvoyants, and she often doesn't even see the outline of the physical body. "A person's physical body can come into it, but rarely because that's seeing the etheric body rather than seeing the aura or the energy field around them. If I'm seeing the etheric, it's usually because it contains leaks or rips that are keeping the aura from being whole. Thus I cannot see it completely. There are only patches of it. It's kind of like a ripped blanket or a torn curtain. Holes in the etheric field are usually the result of trauma, injury, illness, or some other kind of devastating experience."

But beyond seeing the etheric, Dryer says that instead of seeing the layers of the aura like tiers of cake piled one on top of the other, she *experiences* them as changing textures and intensities of visual sensation. She compares this to being immersed in the ocean and feeling water of different temperatures wash by. "Rather than getting into rigid concepts like layers, I tend to see the energy field in terms of movements and waves of energy," she says. "It's as if my vision is telescoping through various levels and dimensions of the energy field, but I don't actually see it neatly arranged in various layers."

This does not mean that Dryer's perception of the human energy field is in any way less detailed than Brennan's. She perceives a dazzling amount of pattern and structure—kaleidoscopic clouds of color shot through with light, complex images, glistening shapes, and gossamer mists. However, not all energy fields are created equal. According to Dryer, shallow people have shallow and humdrum auras. Conversely, the more complex the person, the more complex and interesting their energy field. "A person's energy field is as individual as their fingerprint. I've never really seen any two that look alike," she says.

Like Brennan, Dryer can diagnose illnesses by looking at a person's aura, and when she chooses she can adjust her vision and see the chakras. But Dryer's special skill is the ability to peer deep into a person's psyche and give them an eerily accurate status report of the weaknesses, strengths, needs, and general health of their emotional, psychological, and spiritual being. So profound are her talents in this area that some have likened a session with Dryer to six months of psychotherapy. Numerous clients have credited her with completely transforming their lives, and her files are filled with glowing letters of thanks.

I, too, can attest to Dryer's abilities. In my first reading with her, and although we were virtual strangers, she proceeded to describe things about me that not even my closest friends know. These were not just vague platitudes, but specific and detailed assessments of my talents, vulnerabilities, and personality dynamics. By the end of the two-hour session I was convinced that Dryer had not been looking at my physical presence, but at the energy construct of my psyche itself. I have also had the privilege of talking with and/or listening to the session recordings of over two dozen of Dryer's clients, and have discovered that, almost without exception, others have found her as accurate and insightful as did I.

Doctors Who See the Human Energy Field

Although the existence of the human energy field is not recognized by the medical orthodox community, it has not been completely ignored by medical practitioners. One medical professional who takes the energy field seriously is neurologist and psychiatrist Shafica Karagulla. Karagulla received her degree of doctor of medicine and surgery from the American University of Beirut, Lebanon, and obtained her training in psychiatry under the well-known psychiatrist Professor Sir David K. Henderson, at the Royal Edinburgh Hospital for Mental and Nervous Disorders. She also spent three and a half years as a research associate to Wilder Penfield, the Canadian neurosurgeon whose landmark studies of memory launched both Lashley and Pribram on their quest.

Karagulla began as a skeptic, but after encountering several individuals who could see auras, and confirming their ability to make accurate medical diagnoses as a result of what they saw, she became a believer. Karagulla calls the faculty to see the human energy field *higher sense perception,* or HSP, and in the 1960s she set out to determine if any members of the medical profession also possessed the ability. She put out various feelers among her friends and colleagues, but at first the going was slow. Even doctors who were said to have the ability were reluctant to meet with her. After being put off repeatedly by one such doctor, she finally made an appointment to see him as a patient.

She entered his office, but instead of allowing him to perform a physical examination to diagnose her condition, she challenged him to use his higher sense perception. Realizing he was cornered, he gave in. "All right, stay where you are," he told her. "Don't tell me anything." Then he scanned her body and gave her a quick run-down of her health, including a description of an internal condition she had that would eventually require surgery, a condition she had secretly already diagnosed. He was "correct in every detail," says Karagulla.[10]

As Karagulla's network of contacts expanded, she met doctor after doctor with similar gifts and describes these encounters in her book *Breakthrough to Creativity*. Most of these physicians were unaware that other individuals existed who possessed similar talents, and felt they were alone and peculiar in this regard. Nonetheless, they invariably described what they were seeing as an "energy field" or a "moving web of frequency" around the body and interpenetrating the body. Some saw chakras, but because they were ignorant of the term, they described them as "vortices of energy at certain points along the spine, connected with or influencing the endocrine system." And almost without exception they kept their abilities a secret out of fear of damaging their professional reputations.

Out of respect for their privacy, Karagulla identifies them in her book by first name only but says they include famous surgeons, Cornell University professors of medicine, heads of departments in large hospitals, and Mayo Clinic physicians. "I was continually surprised to find how many members of the medical profession had HSP abilities," she writes. "Most of them felt a little uneasy about their gifts, but finding them useful in diagnosis, they used them. They came from many parts of the country, and although they were unknown to each other, they all reported similar types of experiences." At the end of her report, she concludes, "When many reliable individuals independently report the same kind of phenomena, it is time science takes cognizance of it."[11]

Not all health professionals are so opposed to going public with their abilities. One such individual is Dr. Dolores Krieger, a professor of nursing at New York University. Krieger became interested in the human energy field after participating in a study of the abilities of Oscar Estebany, a well-known Hungarian healer. After discovering that Estebany could raise the hemoglobin levels in ill patients simply by manipulating their fields, Krieger set out to learn more about the mysterious energies involved. She immersed herself in a study of *prana*, the chakras, and the aura, and eventually became a student of Dora Kunz, another well-known clairvoyant. Under Kunz's guidance, she learned how to feel blockages in the human energy field and to heal by manipulating the field with her hands.

Realizing the enormous medical potential of Kunz's techniques, Krieger decided to teach what she had learned to others. Because she knew terms such as *aura* and *chakra* would have negative connotations for many health-care professionals, she decided to call her healing method "therapeutic touch." The first class she taught on therapeutic touch was a master's level course for nurses at New York University entitled "Frontiers in Nursing: The Actualization of Potential for Therapeutic Field Interaction." Both the course and the technique proved so successful that Krieger has since taught therapeutic touch to literally thousands of nurses, and it is now used in hospitals around the world.

The effectiveness of therapeutic touch has also been demonstrated in several studies. For example, Dr. Janet Quinn, an associate professor and assistant director of nursing research at the University of South Carolina at Columbia, decided to see if therapeutic touch could lower the anxiety levels of heart patients. To accomplish this she devised a double-blind study in which one group of nurses trained in the technique would pass their hands over a group of heart patients' bodies. A second group with no training would pass their hands over the bodies of another group of heart patients, but without actually performing the technique. Quinn found that the anxiety levels in the authentically treated patients dropped 17 percent after only five minutes of therapy, but there was no change in anxiety levels among the patients who received the "fake" treatment. Quinn's study was the lead story in the *Science Times* section of the March 26, 1985, issue of the *New York Times*.

Another health professional who lectures widely about the human energy field is University of Southern California heart and lung specialist W. Brugh Joy. Joy, who is a graduate of both Johns Hopkins and the Mayo Clinic, discovered his gift in 1972 while examining a patient in his office. Instead of seeing the aura, Joy initially was only able to feel its presence with his hands. "I was examining a healthy male in his early twenties," he says. "As my hand passed over the solar plexus area, the pit of the stomach, I sensed something that felt like a warm cloud. It seemed to radiate out three to four feet from the body, perpendicular to the surface and to be shaped like a cylinder about four inches in diameter."[12]

Joy went on to discover that all his patients had palpable cylinderlike radiations emanating not only from their stomachs, but from various other points on their bodies. It wasn't until he read an ancient Hindu book about the human energy system that he found he had discovered, or rather *re*discovered, the chakras. Like Brennan, Joy thinks the holographic model offers the best explanation for understanding the human energy field. He also feels that the ability to see auras is latent in all of us. "I believe that reaching expanded states of consciousness is merely the attuning of our central nervous system to perceptive states that have always existed in us but have been blocked by our outer mental conditioning," says Joy.[13]

To prove his point, Joy now spends most of his time teaching others how to sense the human energy field. One of Joy's students is Michael Crichton, the author of such bestsellers as *The Andromeda Strain* and *Sphere,* and the director of the motion pictures *Coma* and *The First Great Train Robbery.* In his recent bestselling autobiography *Travels,* Crichton, who obtained his medical degree from the Harvard University Medical School, describes how he learned to feel and eventually see the human energy field by studying under both Joy and other gifted teachers. The experience astonished and transformed Crichton. "There isn't any delusion. It is absolutely clear that this body energy is a genuine phenomenon of some kind," he states.[14]

Chaos Holographic Patterns

The increasing willingness of doctors to go public with such abilities is not the only change that has taken place since Karagulla did her investigations. Over the past twenty years Valerie Hunt, a physical therapist and professor of kinesiology at UCLA, has developed a way to confirm experimentally the existence of the human energy field. Medical science has long known that humans are electromagnetic beings. Doctors routinely use electrocardiographs to make electrocardiograms (EKGs) or records, of the electrical activity of the heart, and electroencephalographs to make electroencephalograms (EEGs) of the brain's electrical activity. Hunt has discovered that an electromyograph, a device used to measure the electrical activity in the muscles, can also pick up the electrical presence of the human energy field.

Although Hunt's original research involved the study of human muscular movement, she became interested in the energy field after encountering a dancer who said she used her own energy field to help her dance. This inspired Hunt to make electromyograms (EMGs) of the electrical activity in the woman's muscles while she danced, and also to study the effect healers had on the electrical activity in the muscles of people being healed. Her research eventually expanded to include individuals who could see the human energy field, and it was here that she made some of her most significant discoveries.

The normal frequency range of the electrical activity in the brain is between 0 and 100 cycles per second (cps), with most of the activity occurring between 0 and 30 cps. Muscle frequency goes up to about 225 cps, and the heart goes up to about 250 cps, but this is where electrical activity associated with biological function drops off. In addition to these, Hunt discovered that the electrodes of the electromyograph could pick up another field of energy radiating from the body, much subtler and smaller in amplitude than the traditionally recognized body electricities but with frequencies that averaged between 100 and 1600 cps, and which sometimes went even higher. Moreover, instead of emanating from the brain, heart, or muscles, the field was strongest in the areas of the body associated with the chakras. "The results were so exciting that I simply was not able to sleep that night," says Hunt. "The scientific model I had subscribed to throughout my life just couldn't explain these findings."[15]

Hunt also discovered that when an aura reader saw a particular color in a person's energy field, the electromyograph always picked up a specific pattern of frequencies that Hunt learned to associate with that color. She was able to see this pattern on an oscilloscope, a device that converts electrical waves into a visual pattern on a monochromatic video display screen. For example, when an aura reader saw blue in a person's energy field, Hunt could confirm that it was blue by looking at the pattern on the oscilloscope. In one experiment she even tested eight aura readers simultaneously to see if they would agree with the oscilloscope as well as with each other. "It was the same right down the line," says Hunt.[16]

Once Hunt confirmed the existence of the human energy field, she, too, became convinced that the holographic idea offers one model for understanding it. In addition to its frequency aspects, she points out that the energy field, and indeed all of the body's electrical systems, is holographic in another way. Like the information in a hologram, these systems are distributed globally throughout the body. For instance, the electrical activity measured by an electroencephalograph is strongest in the brain, but an EEG reading can also be made by attaching an electrode to the toe. Similarly, an EKG can be picked up in the little finger. It's stronger and higher in amplitude in the heart, but its frequency and pattern are the same everywhere in the body. Hunt believes this is significant. Although every portion of what she calls the "holographic field reality" of the aura contains aspects of the whole energy field, different portions are not absolutely identical to each other. These differing amplitudes keep the energy field from being a static hologram, and instead allow it to be dynamic and flowing, says Hunt.

One of Hunt's most startling findings is that certain talents and abilities seem to be related to the presence of specific frequencies in a person's energy field. She has found that when the main focus of a person's consciousness is on the material world, the frequencies of their energy field tend to be in the lower range and are not too far removed from the 250 cps of the body's biological frequencies. In addition to these, people who are psychic or who have healing abilities also have frequencies of roughly 400 to 800 cps in their field. People who can go into trance and apparently channel other information sources through them, skip these "psychic" frequencies entirely and operate in a narrow band between 800 and 900 cps. "They don't have any psychic breadth at all," states Hunt. "They're up there in their own field. It's narrow. It's pinpointed, and they literally are almost out of it."[17]

People who have frequencies above 900 cps are what Hunt calls mystical personalities. Whereas psychics and trance mediums are often just conduits of information, mystics possess the wisdom to know what to do with the information, says Hunt. They are aware of the cosmic interrelatedness of all things and are in touch with every level of human experience. They are anchored in ordinary reality, but often have both psychic and trance abilities. However, their frequencies also extend way beyond the bands associated with these capabilities. Using a modified electromyogram (an electromyogram can normally detect frequencies only up to 20,000 cps) Hunt has encountered individuals who have frequencies as high as 200,000 cps in their energy fields. This is intriguing, for mystical traditions have often referred to highly spiritual individuals as possessing a "higher vibration" than normal people. If Hunt's findings are correct, they seem to add credence to this assertion.

Another of Hunt's discoveries involves the new science of chaos. As its name implies, chaos is the study of chaotic phenomena, i.e., processes that are so haphazard they do not appear to be governed by any laws. For example, when smoke rises from an extinguished candle it flows upward in a thin and narrow stream. Eventually the structure of the stream breaks down and becomes turbulent. Turbulent smoke is said to be chaotic because its behavior can no longer be predicted by science. Other examples of chaotic phenomena include water when it crashes at the bottom of a waterfall, the seemingly random electrical fluctuations that rage through the brain of an epileptic during a seizure, and the weather when several different temperature and air-pressure fronts collide.

In the past decade science has discovered that many chaotic phenomena are not as disordered as they seem and often contain hidden patterns and regularities (recall Bohm's assertion that there is no such thing as disorder, only orders of indefinitely high degree). Scientists have also discovered mathematical ways of finding some of the regularities that lie hidden in chaotic phenomena. One of these involves a special kind of mathematical analysis that can convert data about a chaotic phenomenon into a shape on a computer screen. If the data contains no hidden patterns, the resulting shape will be a straight line. But if the chaotic phenomenon does contain hidden regularities, the shape on the computer screen will look something like the spiral designs children make by winding colored yarn around an array of nails pounded into a board. These shapes are called "chaos patterns" or "strange attractors" (because the lines that compose the shape seem to be attracted again and again to certain areas of the computer screen, just as the yarn might be said to be repeatedly "attracted" to the array of nails around which it is wound).

When Hunt observed energy field data on the oscilloscope, she noticed that it changed constantly. Sometimes it came in great clumps, sometimes it waned and became patchy, as if the energy field itself were in an unceasing state of fluctuation. At first glance these changes seemed random, but Hunt sensed intuitively they possessed some order. Realizing that chaos analysis might reveal whether she was right or not, she sought out a mathematician. First they ran four seconds of data from an EKG through the computer to see what would happen. They got a straight line. Then they ran the same amount of data from an EEG and an EMG. The EEG produced a straight line and the EMG produced a slightly swollen line, but still no chaos pattern. Even when they submitted data from the lower frequencies of the human energy field, they got a straight line. But when they analyzed the very high frequencies of the field they met with success. "We got the most dynamic chaos pattern you ever saw," says Hunt.[18]

This meant that although the kaleidoscopic changes taking place in the energy field appeared to be random, they were actually very highly ordered and rich with pattern. "The pattern is never a repeatable one, but it's so dynamic and complex, I call it a chaos holograph pattern," Hunt states.[19]

Hunt believes her discovery was the first true chaos pattern to be found in a major electrobiological system. Recently researchers have found chaos patterns in EEG recordings of the brain, but they needed many minutes of data from numerous electrodes to obtain such a pattern. Hunt obtained a chaos pattern from three to four seconds of data recorded by one electrode, suggesting that the human energy field is far richer in information and possesses a far more complex and dynamic organization than even the electrical activity of the brain.

What Is the Human Energy Field Made Of?

Despite the human energy field's electrical aspects, Hunt does not believe it is purely electromagnetic in nature. "We have a feeling that it is much more complex and without doubt composed of an as yet undiscovered energy," she says.[20]

What is this undiscovered energy? At present we do not know. Our best clue comes from the fact that almost without exception psychics describe it as having a higher frequency or vibration than normal matter-energy. Given the uncanny accuracy talented psychics have in perceiving illnesses in the energy field, we should perhaps pay serious attention to this observation. The universality of this perception— even ancient Hindu literature asserts that the energy body possesses a higher vibration than normal matter—may be an indication that such individuals are intuiting an important fact about the energy field.

Ancient Hindu literature also describes matter as being composed of *anu,* or "atoms," and says that the subtle vibratory energies of the human energy field exist *paramanu,* or literally "beyond the atom." This is interesting, for Bohm also believes that at a subquantum level *beyond the atom* there are many subtle energies still unknown to science. He confesses that he does not know whether the human energy field exists or not, but in commenting on the possibility, he states, "The implicate order has many levels of subtlety. If our attention can go to those levels of subtlety, then we should be able to see more than we ordinarily see."[21]

It is worth noting that we really don't know what *any* field is. As Bohm has said, "What is an electric field? We don't know."[22] When we discover a new kind of field it seems mysterious. Then we name it, get used to dealing with it and describing its properties, and it no longer seems mysterious. But we still do not know what an electric or a gravitational field really is. As we saw in an earlier chapter, we don't even know what electrons are. We can only describe how they behave. This suggests that the human energy field will also ultimately be defined in terms of how it behaves, and research such as Hunt's will only further our understanding.

Three-Dimensional Images in the Aura

If these inordinately subtle energies are the stuff from which the human energy field is made, we may rest assured that they possess qualities unlike the kinds of energy with which we are normally familiar. One of these is evident in the human energy field's nonlocal characteristics. Another, and one that is particularly holographic, is the aura's ability to manifest as an amorphous blur of energy, or occasionally form itself into three-dimensional images. Talented psychics often report seeing such "holograms" floating in people's auras. These images are usually of objects and ideas that hold a prominent position in the thoughts of the person around whom they are seen. Some occult traditions hold that such images are a product of the third, or mental, layer of the aura, but until we have the means to confirm or deny this allegation, we must confine ourselves to the experiences of the psychics who are able to see images in the aura.

One such psychic is Beatrice Rich. As often happens, Rich's powers manifested at an early age. When she was a child, objects in her presence would occasionally move about on their own accord. When she grew older she discovered she knew things about people she had no normal means of knowing. Although she began her career as an artist, her clairvoyant talents proved so impressive that she decided to become a full-time psychic. Now she gives readings for individuals from all walks of life, from housewives to chief executives of corporations, and articles about her work have appeared in such diverse publications as *New York* magazine, *World Tennis,* and *New York Woman.*

Rich often sees images floating around or hovering near her clients. Once she saw silver spoons, silver plates, and similar objects circling around a man's head. Because it was early in her explorations of psychic phenomena, the experience startled her. At first she did not know why she was seeing what she was seeing. But finally she told the man and discovered that he was in the import/export business and traded in the very objects she was seeing circling his head. The experience was riveting and changed her perceptions forever.

Dryer has had many similar experiences. Once during a reading she saw a bunch of potatoes whirling around a woman's head. Like Rich, she was at first dumbfounded but summoned her courage and asked the woman if potatoes had any special meaning for her. The woman laughed and handed Dryer her business card. "She was from the Idaho Potato Board, or something like that," says Dryer. "You know, the potato grower's equivalent of the American Dairy Association."[23]

These images don't always just hover in the aura, but sometimes can appear to be ghostly extensions of the body itself. On one occasion Dryer saw a wispy and holographiclike layer of mud clinging to a woman's hands and arms. Given the woman's impeccable grooming and expensive attire, Dryer could not imagine why thoughts of mucking around in some kind of viscous sludge would be occupying her mind. Dryer asked her if she understood the image, and the woman nodded, explaining that she was a sculptor and had tried out a new medium that morning that had clung to her arms and hands exactly as Dryer had described.

I, too, have had similar experiences when looking at the energy field. Once, while deep in thought about a novel I was working on about werewolves (as some readers may be aware, I have a fondness for writing fiction about folkloric subjects), I noticed that the ghostly image of a werewolf's body had formed around my own body. I would quickly like to stress that this was a purely visual phenomenon and at no time did I feel I had in any way become a werewolf. Nonetheless, the holographiclike image that enveloped my body was real enough that when I lifted my arm I could actually see the individual hairs in the fur and the way the canine nails protruded from the wolfish hand that encased my own hand. Indeed, everything about these features was absolutely real, save that they were translucent and I could see my own flesh-and-blood hand beneath them. The experience should have been frightening, but for some reason it wasn't, and I found myself only fascinated by what I was seeing.

What was significant about this experience was that Dryer was my house guest at the time and happened to walk into the room while I was still sheathed in this phantomlike werewolf body. She reacted immediately and said, "Oh my, you must be thinking about your were-wolf novel because you've become a werewolf." We compared notes and discovered that we were each observing the same features. We became involved in conversation, and as my thoughts strayed from the novel, the werewolf image slowly faded.

Movies in the Aura

The images that psychics see in the energy field are not always static. Rich says she often sees what looks like a little transparent movie going on around a client's head: "Sometimes I see a small image of the person behind their head or shoulders doing various things they do in life. My clients tell me that my descriptions are very accurate and specific. I can see their offices and what their bosses look like. I can see what they've thought of and what's happened to them during the last six months. Recently I told a client that I could see her home and she had masks and flutes hanging on her wall. She said, 'No, no, no.' I said yes, there are musical instruments hanging on the wall, mostly flutes, and there are masks. And then she said, 'Oh, that's my summer home.' "[24]

Dryer says she also sees what look like three-dimensional movies in a person's energy field. "Usually they're in color, but they can also be brown, or look like tintypes. Often they depict a story about the person that can take anywhere from five minutes to an hour to unfold. The images are also incredibly detailed. When I see a person sitting in a room I can tell them how many plants are in the room, how many leaves are on each plant, and how many bricks are in the wall. I usually don't get into such minute description unless it seems pertinent."[25]

I can attest to Dryer's accuracy. I have always been an organized person, and when I was a child I was quite precocious in this regard. Once when I was five years old I spent several hours meticulously storing and organizing all of my toys in a closet. When I was finished I showed my mother what I had done and admonished her please not to touch anything in the closet because I did not want her messing up my carefully ordered arrangements. My mother's account of this incident has amused the family ever since. During my first reading with Dryer she described this incident in detail, as well as many other events in my life, as she watched it unfold like a movie in my energy field. She, too, chuckled as she described it.

Dryer likens the images she sees to holograms and says that when she chooses one and starts to watch it, it seems to expand and fill the entire room. "If I see something going on with a person's shoulder, such as an injury, suddenly the whole scene widens. That's when I get the sense that it's a hologram because sometimes I feel I can step right into it and be a part of it. It's not happening to me, but around me. It's almost as if I'm in a three-dimensional movie, a holographic movie, with the person."[26]

Dryer's holographic vision is not limited to events from a person's life. She sees visual representations of the operations of the unconscious mind as well. As we all know, the unconscious mind speaks in a language of symbols and metaphors. This is why dreams often seem so nonsensical and mysterious. However, once one learns how to interpret the language of the unconscious, the meanings of dreams become clear. Dreams are not the only things that are written in the parlance of the unconscious. Individuals who are familiar with the language of the psyche—a language psychologist Erich Fromm calls the "forgotten language," because most of us have forgotten how to interpret it—recognize its presence in other human creations such as myths, fairy tales, and religious visions.

Some of the holographic movies Dryer sees in the human energy field are also written in this language and resemble the metaphorical messages of dreams. We now know that the unconscious mind is active not only while we dream, but all of the time. Dryer is able to peel back a person's waking self and gaze directly at the unceasing river of images that is always flowing through their unconscious mind. And both practice and her natural, intuitive gifts have made her extremely skilled at deciphering the language of the unconscious. "Jungian psychologists love me," says Dryer.

In addition, Dryer has a special way of knowing whether she has interpreted an image correctly. "If I haven't explained it correctly, it doesn't go away," she states. "It just stays in the energy field. But once I've told the person everything they need to know about a particular image, it begins to dissolve and disappear."[27] Dryer thinks this is because it is a client's own unconscious mind that chooses what images to show her. Like Ullman, she believes the psyche is always trying to teach the conscious self things it needs to know to become healthier and happier, and to grow spiritually.

Dryer's ability to observe and interpret the innermost workings of a person's psyche is one of the reasons she is able to effect such profound transformations in many of her clients. The first time she described the stream of images she saw unfolding in my own energy field, I had the uncanny sensation she was telling me about one of my own dreams, save that it was a dream I had not yet dreamed. At first the phantasmagoria of images was only mysteriously familiar, but as she unraveled and explained each symbol and metaphor in turn, I recognized the machinations of my inner self, both the things I accepted and the things I was less willing to embrace. Indeed, it is clear from the work of psychics like Rich and Dryer that there is an enormous amount of information in the energy field. One wonders if this is perhaps why Hunt obtained such a pronounced chaos pattern when she analyzed data from the aura.

The ability to see images in the human energy field is not new. Nearly three hundred years ago the great Swedish mystic Emanuel Swedenborg reported that he could see a "wave-substance" around people, and in the wave-substance a person's thoughts were visible as images he called "portrayals." In commenting on the inability of other people to see this wave-substance around the body, he observed, "I could see solid concepts of thought as though they were surrounded by a kind of wave. But nothing reaches [normal] human sensation except what is in the middle and seems solid."[28] Swedenborg could also see portrayals in his own energy field: "When I was thinking about someone I knew, then his image appeared as he looked when he was named in human presence; but all around, like something flowing in waves, was everything I had known and thought about him from boyhood."[29]

Holographic Body Assessment

Frequency is not the only thing that is distributed holographically throughout the field. Psychics report that the wealth of personal information the field contains can also be found in every portion of the body's aura. As Brennan puts it, "The aura not only represents, but also contains, the whole."[30] California clinical psychologist Ronald Wong Jue agrees. Jue, a former president of the Association for Transpersonal Psychology and a talented clairvoyant, has found that an individual's history is even contained in the "energy patterns" inherent *in* the body. "The body is a kind of microcosm, a universe unto itself reflecting all of the different factors that a person is dealing with and trying to integrate," says Jue.

Like Dryer and Rich, Jue has the psychic ability to tune into movies about the important issues in a person's life, but instead of seeing them in the energy field, he conjures them up in his mind's eye by laying his hands on a person and literally psychometrizing their body. Jue says this technique enables him to determine quickly the emotional scripts, core issues, and relationship patterns that are most prominent in a person's life, and often uses it on his patients to facilitate the therapeutic process. "The technique was actually taught to me by a psychiatrist colleague of mine named Ernest Pecci," Jue states. "He called it 'body reading.' Instead of talking about the etheric body and things like that, I chose to use the holographic model as a way of explaining it and call it Holographic Body Assessment."[31] In addition to using it in his clinical practice, Jue also gives seminars in which he teaches others how to use the technique.

X-Ray Vision

In the last chapter we explored the possibility that the body is not a solid construct, but is itself a kind of holographic image. Another faculty possessed by many clairvoyants seems to support this notion, that is, the ability to literally look inside a person's body. Individuals who are gifted at seeing the energy field can also often adjust their vision and see through the flesh and bones of the body as if they were no more than layers of colored mist.

During the course of her research, Karagulla discovered a number of people, both in and out of the medical profession, who possessed this X-ray vision. One, a woman she identifies as Diane, was the head of a corporation. Just before meeting Diane, Karagulla wrote, "For me as a psychiatrist to be meeting somebody who was reported to be able to 'see' right through me was a shattering reversal of my usual procedures."[32]

Karagulla put Diane through a lengthy series of tests, introducing her to people and having her make on-the-spot diagnoses. On one of these occasions Diane described a woman's energy field as "wilted" and "broken into fragments" and said this indicated she had a serious problem in her physical body. She then looked into the woman's body and saw that there was an intestinal blockage near her spleen. This surprised Karagulla because the woman showed none of the symptoms that usually indicated such a serious condition. Nonetheless, the woman went to her doctor, and X rays revealed a blockage in the precise area Diane had described. Three days later the woman underwent surgery to have the life-threatening obstruction removed.

In another series of tests Karagulla had Diane diagnose patients at random in the outpatient clinic of a large New York hospital. After Diane made a diagnosis Karagulla would determine the accuracy of her observations by referring to the patient's records. On one of these occasions Diane looked at a patient unknown to both of them and told Karagulla that the woman's pituitary gland (a gland deep in the brain) was missing, her pancreas looked as if it was not functioning properly, her breasts had been affected but were now missing, she didn't have enough energy going through her spine from the waist down, and she had trouble with her legs. The medical report on the woman revealed that her pituitary gland had been surgically removed, she was taking hormones which affected her pancreas, she had had a double mastectomy due to cancer, an operation on her back to decompress her spinal cord and relieve pains in her legs, and her nerves had been damaged, making it difficult for her to empty her bladder.

In case after case Diane revealed that she could gaze effortlessly into the depths of the physical body. She gave detailed descriptions of the condition of the internal organs. She saw the state of the intestines, the presence or absence of the various glands, and even described the density or brittleness of the bones. Concludes Karagulla, "Although I could not evaluate her findings regarding the energy body, her observations of physical conditions correlated with amazing accuracy with the medical diagnoses."[33]

Brennan is also skilled at looking into the human body and calls the ability "internal vision." Using internal vision she has accurately diagnosed a wide range of conditions including bone fractures, fibroid tumors, and cancer. She says she can often tell the condition of an organ by its color: for example, a healthy liver looks dark red, a jaundiced liver looks a sickly yellow-brown, and the liver of an individual undergoing chemotherapy usually looks green-brown. Like many other psychics with internal vision, Brennan can adjust the focus of her vision and even see microscopic structures, such as viruses and individual blood cells.

I have personally encountered several psychics with internal vision and can corroborate its authenticity. One psychic I have seen demonstrate the ability is Dryer. On one of these occasions she not only accurately diagnosed an internal medical problem I was having, but offered some startling information of an entirely different nature along with it. A few years back I started having trouble with my spleen. To try and remedy the situation, I began performing daily visualization exercises, seeing images of my spleen in a state of wholeness and health, seeing it being bathed in healing light, and so on. Unfortunately, I am a very impatient person, and when I did not have overnight success I got angry. During my next meditation I mentally scolded my spleen and warned it in no uncertain terms that it better start doing what I wanted. This incident took place purely in the privacy of my own thoughts, and I quickly forgot about it.

A few days later I saw Dryer and asked her if she could look into my body and tell me if there was anything I should be aware of (I did not tell her about my health problem). Nonetheless, she immediately described what was wrong with my spleen and then paused, scowling as if she was confused. "Your spleen's very upset about something," she murmured. And then suddenly it hit her. "Have you been *yelling* at your spleen?" I sheepishly admitted that I had. Dryer all but threw her hands up. "You mustn't do that. Your spleen became ill because it thought it was doing what you wanted. That was because you were unconsciously giving it the wrong directions. Now that you've yelled at it, it's really confused." She shook her head with concern. "Never, never get angry at your body or your internal organs," she advised. "Only send them positive messages."

The incident not only revealed Dryer's skill at looking inside the human body, but also seemed to suggest that my spleen has some sort of mentality or consciousness all of its own. It reminded me not only of Pert's assertion that she no longer knows where the brain leaves off and the body begins, but made me wonder if perhaps all of the body's subcomponents—glands, bones, organs, and cells—possess their own intelligence? If the body is truly holographic, it may be that Pert's remark is more correct than we realize, and the consciousness of the whole is very much contained in all of its parts.

Internal Vision and Shamanism

In some shamanic cultures internal vision is one of the prerequisites for becoming a shaman. Among the Araucanian Indians of Chile and the Argentine pampas, a newly initiated shaman is taught to pray specifically for the faculty. This is because the shaman's major role in Araucanian culture is to diagnose and heal illness, for which internal vision is considered essential.[34] Australian shamans refer to the ability as the "strong eye," or "seeing with the heart."[35] The Jivaro Indians of the forested eastern slopes of the Ecuadorian Andes acquire the ability by drinking an extract of a jungle vine called *ayahuasca,* a plant containing a hallucinogenic substance believed to bestow psychic abilities on the imbiber. According to Michael Harner, an anthropologist at the New School for Social Research in New York who specializes in shamanic studies, *ayahuasca* permits the Jivaro shaman "to see into the body of the patient as though it were glass."[36]

Indeed, the ability to "see" an illness—whether it involves actually looking inside the body or seeing the malady represented as a kind of metaphorical hologram, such as a three-dimensional image of a demonic and repulsive creature inside or near the body—is universal in shamanic traditions. But whatever the culture in which internal vision is reported, its implications are the same. The body is an energy construct and ultimately may be no more substantive than the energy field in which it is embedded.

The Energy Field as Cosmic Blueprint

The idea that the physical body is just one more level of density in the human energy field and is itself a kind of hologram that has coalesced out of the interference patterns of the aura may explain both the extraordinary healing powers of the mind and the enormous control it has over the body in general. Because an illness can appear in the energy field weeks and even months before it appears in the body, many psychics believe that disease actually originates in the energy field. This suggests that the field is in some way more primary than the physical body and functions as a kind of blueprint from which the body gets its structural cues. Put another way, the energy field may be the body's own version of an implicate order.

This may explain Achterberg's and Siegel's findings that patients are already "imaging" their illnesses many months before the illnesses manifest in their bodies. At present, medical science is at a loss to explain how mental imagery could actually create an illness. But, as we have seen, ideas that are prominent in our thoughts quickly appear as images in the energy field. If the energy field is the blueprint that guides and molds the body, it may be that by imaging an illness, even unconsciously, and repeatedly reinforcing its presence in the field, we are in effect programming the body to manifest the illness.

Similarly, this same dynamic linkage between mental images, the energy field, and the physical body may be one of the reasons imagery and visualization can also heal the body. It may even help explain how faith and meditation on religious images enable stigmatists to grow nail-like fleshy protuberances from their hands. Our current scientific understanding is at a loss to explain such a biological capacity, but again, constant prayer and meditation may cause such images to become so impressed in the energy field that the constant repetition of these patterns is finally given form in the body.

One researcher who believes it is the energy field that molds the body and not the other way around is Richard Gerber, a Detroit physician who has spent the last twelve years investigating the medical implications of the body's subtle energy fields. "The etheric body is a holographic energy template that guides the growth and development of the physical body," says Gerber.[37]

Gerber believes that the distinct layers some psychics see in the aura also play a factor in the dynamic relationship among thought, the energy field, and the physical body. Just as the physical body is subordinate to the etheric, the etheric body is subordinate to the astral/emotional body, the astral/emotional to the mental, and so on, says Gerber, with each body functioning as the template for the one before it. Thus the subtler the layer of the energy field in which an image or thought manifests, the greater its ability to heal and reshape the body. "Because the mental body feeds energy into the astral/emotional body, which then funnels down into the etheric and physical bodies, healing a person at the mental level is stronger and produces longer lasting results than healing from either the astral or etheric levels," says Gerber.[38]

Physicist Tiller agrees. "The thoughts that one creates generate patterns at the mind level of nature. So we see that illness, in fact, eventually becomes manifest from the altered mind patterns through the rachet effect—first, to effects at the etheric level and then, ultimately, at the physical level [where] we see it openly as disease." Tiller believes the reason illnesses often recur is that medicine currently treats only the physical level. He feels that if doctors could treat the energy field as well, they would bring about longer lasting cures. Until then, many treatments "will not be permanent because we have not altered the basic hologram at the mind and spiritual levels," he states.[39]

In a wide-ranging speculation Tiller even suggests that the universe itself started as a subtle energy field and gradually became dense and material through a similar rachet effect. As he sees it, it may be that God created the universe as a divine pattern or idea. Like the image a psychic sees floating in the human energy field, this divine pattern functioned as a template, influencing and molding increasingly less subtle levels of the cosmic energy field "on down the line via a series of holograms," until it eventually coalesced into a hologram of a physical universe.[40]

If this is true, it suggests that the human body is holographic in another way, for each of us truly would be a universe in miniature. Furthermore, if our thoughts can cause ghostly holographic images to form, not only in our own energy fields, but in the subtle energetic levels of reality itself, it may help explain how the human mind is able to effect some of the miracles we examined in the previous chapter. It may even explain synchronicities, or how processes and images from the innermost depths of our psyche manage to take form in external reality. Again, it may be that our thoughts are constantly affecting the subtle energetic levels of the holographic universe, but only emotionally powerful thoughts, such as the ones that accompany moments of crisis and transformation—the kind of events that seem to engender synchronicities—are potent enough to manifest as a series of coincidences in physical reality.

A Participatory Reality

Of course, these processes are not contingent on the subtle energy fields of the universe being stratified into rigidly defined layers. They could also work even if the subtle fields of the universe are a smooth continuum. In fact, given how sensitive these subtle fields are to our thoughts, we must be very careful when trying to form set ideas about their organization and structure. What we believe about them may in fact help mold and create their structure.

This is perhaps why psychics disagree about whether the human energy field is divided into layers. Psychics who believe in clearly defined layers may actually be causing the energy field to form itself into layers. The individual whose energy field is being observed may also participate in this process. Brennan is very frank about this and notes that the more one of her clients understands the difference between the layers, the clearer and more distinct the layers of their energy field become. She admits that the structure she sees in the energy field is thus but one system, and others have come up with other systems. For example, the authors of the tantras, a collection of Hindu yogic texts written during the fourth through sixth centuries A.D., perceived only three layers in the energy field.

There is evidence that the structures clairvoyants inadvertently create in the energy field can be remarkably long-lived. For centuries the ancient Hindus believed that each chakra also had a Sanskrit letter written in its center. Japanese researcher Hiroshi Motoyama, a clinical psychologist who has successfully developed a technique for measuring the electrical presence of the chakras, says that he first became interested in the chakras because his mother, a simple woman with natural clairvoyant gifts, could see them clearly. However, for years she was puzzled because she could see what looked like an inverted sailboat in her heart chakra. It wasn't until Motoyama began his own investigations that he discovered what his mother was seeing was the Sanskrit letter *yam*, the letter the ancient Hindus perceived in the heart chakra.[41] Some psychics, such as Dryer, say that they also see Sanskrit letters in the chakras. Others do not. The only explanation appears to be that psychics who see the letters are actually tuning into holographic structures long ago imposed on the energy field by the beliefs of the ancient Hindus.

At first glance this notion may seem strange, but it does have a precedent. As we have seen, one of the basic tenets of quantum physics is that we are not discovering reality, but participating in its creation. It may be that as we probe deeper into the levels of reality beyond the atom, the levels where the subtle energies of the human aura appear to lie, the participatory nature of reality becomes even more pronounced. Thus we must be extremely cautious about saying that we have discovered a particular structure or pattern in the human energy field, when we may have actually created what we have found.

Mind and the Human Energy Field

It is significant that an examination of the human energy field leads one to precisely the same conclusion Pribram made after discovering that the brain converts sensory import into a language of frequencies. That is, that we have two realities: one in which our bodies appear to be concrete and possess a precise location in space and time; and one in which our very being appears to exist primarily as a shimmering cloud of energy whose ultimate location in space is somewhat ambiguous. This realization brings with it some profound questions. One is, what becomes of mind? We have been taught that our mind is a product of our brain, but if the brain and the physical body are just holograms, the densest part of an increasingly subtle continuum of energy fields, what does this say about the mind? Human energy field research provides an answer.

Recently a discovery made by neurophysiologists Benjamin Libet and Bertram Feinstein at Mount Zion Hospital in San Francisco has been causing a stir in the scientific community. Libet and Feinstein measured the time it took for a touch stimulus on a patient's skin to reach the brain as an electrical signal. The patient was also asked to push a button when he or she became aware of being touched. Libet and Feinstein found that the brain registered the stimulus in 0.0001 of a second after it occurred, and the patient pressed the button 0.1 of a second after the stimulus was applied.

But, remarkably, the patient didn't report being consciously aware of either the stimulus or pressing the button *for almost 0.5 second.* This meant that the decision to respond was being made by the patient's unconscious mind. The patient's awareness of the action was the slow man in the race. Even more disturbing, none of the patients

Libet and Feinstein tested were aware that their unconscious minds had already caused them to push the button before they had consciously decided to do so. Somehow their brains were creating the comforting delusion that they had consciously controlled the action even though they had not.[42] This has caused some researchers to wonder if free will is an illusion. Later studies have shown that one and a half seconds before we "decide" to move one of our muscles, such as lift a finger, our brain has already started to generate the signals necessary to accomplish the movement.[43] Again, who is making the decision, the conscious mind or the unconscious mind?

Hunt does such findings one better. She has discovered that the human energy field responds to stimuli even before the brain does. She has taken EMG readings of the energy field and EEG readings of the brain simultaneously and discovered that when she makes a loud sound or flashes a bright light, the EMG of the energy field registers the stimulus before it ever shows up on the EEG. What does it mean? "I think we have way overrated the brain as the active ingredient in the relationship of a human to the world," says Hunt. "It's just a real good computer. But the aspects of the mind that have to do with creativity, imagination, spirituality, and all those things, I don't see them in the brain at all. The mind's not in the brain. It's in that darn field."[44]

Dryer has also noticed that the energy field responds before a person consciously registers a response. As a consequence, instead of trying to judge her client's reactions by watching their facial expressions, she keeps her eyes closed and watches how their energy fields react. "As I speak I can see the colors change in their energy field. I can see how they feel about what I'm saying without having to ask them. For instance, if their field becomes foggy I know they're not understanding what I'm telling them," she states.[45]

If the mind is not in the brain, but in the energy field that permeates both the brain and the physical body, this may explain why psychics such as Dryer see so much of the content of a person's psyche in the field. It may also explain how my spleen, an organ not normally associated with thought, managed to have its own rudimentary form of intelligence. Indeed, if the mind is in the field, it suggests that our awareness, the thinking, feeling part of ourselves, may not even be confined to the physical body, and as we will see, there is considerable evidence to support this idea as well.

But first we must turn our attention to another issue. The solidity of the body is not the only thing that is illusory in a holographic universe. As we have seen, Bohm believes that even time itself is not absolute, but unfolds out of the implicate order. This suggests that the linear division of time into past, present, and future is also just another construct of the mind. In the next chapter we will examine the evidence that supports this idea as well as the ramifications this view has for our lives in the here and now.

PART III

SPACE AND TIME

Shamanism and similar mysterious areas of research have gained in significance because they postulate new ideas about mind and spirit. They speak of things like vastly expanding the realm of consciousness . . . the belief, the knowledge, and even the experience that our physical world of the senses is a mere illusion, a world of shadows, and that the three-dimensional tool we call our body serves only as a container or dwelling place for Something infinitely greater and more comprehensive than that body and which constitutes the matrix of the real life.

—Holger Kalweit
Dreamtime and Inner Space

7

Time Out of Mind

The "home" of the mind, as of all things, is the implicate order.
At this level, which is the fundamental plenum for the entire
manifest universe, there is no linear time. The implicate domain
is atemporal; moments are not strung together serially like
beads on a string.

—Larry Dossey
Recovering the Soul

As the man gazed off into space, the room he was in became ghostly
and transparent, and in its place materialized a scene from the distant
past. Suddenly he was in the courtyard of a palace, and before him was
a young woman, olive-skinned and very pretty. He could see her gold
jewelry around her neck, wrists, and ankles, her white translucent
dress, and her black braided hair gathered regally under a high
square-shaped tiara. As he looked at her, information about her life
flooded his mind. He knew she was Egyptian, the daughter of a prince,
but not a pharaoh. She was married. Her husband was slender and
wore his hair in a multitude of small braids that fell down on both sides
of his face.

The man could also fast-forward the scene, rushing through the
events of the woman's life as if they were no more than a movie. He
saw that she died in childbirth. He watched the lengthy and intricate
steps of her embalming, her funeral procession, the rituals that accom-

panied her being placed in her sarcophagus, and when he finished, the images faded and the room once again came back into view.

The man's name was Stefan Ossowiecki, a Russian-born Pole and one of the century's most gifted clairvoyants, and the date was February 14, 1935. His vision of the past had been evoked when he handled a fragment of a petrified human foot.

Ossowiecki proved so adept at psychometrizing artifacts that he eventually came to the attention of Stanislaw Poniatowski, a professor at the University of Warsaw and the most eminent ethnologist in Poland at the time. Poniatowski tested Ossowiecki with a variety of flints and other stone tools obtained from archaeological sites around the world. Most of these *lithics*, as they are called, were so nondescript that only a trained eye could tell they had been shaped by human hands. They were also precertified by experts so that Poniatowski knew their ages and historical origins, information he kept carefully concealed from Ossowiecki.

It did not matter. Again and again Ossowiecki identified the objects correctly, describing their age, the culture that had produced them, and the geographical locations where they had been found. On several occasions the locations Ossowiecki cited disagreed with the information Poniatowski had written in his notes, but Poniatowski discovered that it was always his notes that were in error, not Ossowiecki's information.

Ossowiecki always worked the same. He would take the object in his hands and concentrate until the room before him, and even his own body, became shadowy and almost nonexistent. After this transition occurred, he would find himself looking at a three-dimensional movie of the past. He could then go anywhere he wanted in the scene and see anything he chose. While he was gazing into the past, Ossowiecki even moved his eyes back and forth as if the things he was describing possessed an actual physical presence before him.

He could see the vegetation, the people, and the dwellings in which they lived. On one occasion, after handling a stone implement from the Magdalenian culture, a Stone Age people who flourished in France about 15,000 to 10,000 B.C., Ossowiecki told Poniatowski that Magdalenian women had very complex hair styles. At the time this seemed absurd, but subsequent discoveries of statues of Magdalenian women with ornate coiffures proved Ossowiecki right.

Over the course of the experiments Ossowiecki offered over one hundred such pieces of information, details about the past that at first seemed inaccurate, but later proved correct. He said that Stone Age peoples used oil lamps and was vindicated when excavations in Dordogne, France, uncovered oils lamps of the exact size and style he described. He made detailed drawings of the animals various peoples hunted, the style of the huts in which they lived, and their burial customs—assertions that were all later confirmed by archaeological discoveries.[1]

Poniatowski's work with Ossowiecki is not unique. Norman Emerson, a professor of anthropology at the University of Toronto and founding vice president of the Canadian Archaeological Association, has also investigated the use of clairvoyants in archaeological work. Emerson's research has centered around a truck driver named George McMullen. Like Ossowiecki, McMullen has the ability to psychometrize objects and use them to tune into scenes from the past. McMullen can also tune into the past simply by visiting an archaeological site. Once there, he paces back and forth until he gets his bearings. Then he begins to describe the people and culture that once flourished at the site. On one such occasion Emerson watched as McMullen bounded over a patch of bare ground, pacing out what he said was the location of an Iroquois longhouse. Emerson marked the area with survey pegs and six months later uncovered the ancient structure exactly where McMullen said it would be.[2]

Although Emerson began as a skeptic, his work with McMullen has made him a believer. In 1973, at an annual conference of Canada's leading archaeologists, he stated, "It is my conviction that I have received knowledge about archaeological artifacts and archaeological sites from a psychic informant who relates this information to me without any evidence of the conscious use of reasoning." He concluded his talk by saying that he felt McMullen's demonstrations opened "a whole new vista" in archaeology, and research into the further use of psychics in archaeological investigations should be given "first priority."[3]

Indeed, *retrocognition*, or the ability of certain individuals to shift the focus of their attention and literally gaze back into the past, has been confirmed repeatedly by researchers. In a series of experiments conducted in the 1960s, W. H. C. Tenhaeff, the director of the Parapsychological Institute of the State University of Utrecht, and Marius Valkhoff, dean of the faculty of arts at the University of Witwatersrand, Johannesburg, South Africa, found that the great Dutch psychic, Gerard Croiset, could psychometrize even the smallest fragment

of bone and accurately describe its past.[4] Dr. Lawrence LeShan, a
New York clinical psychologist, and another skeptic-turned-believer,
has conducted similar experiments with the noted American psychic,
Eileen Garrett.[5] At the 1961 annual meeting of the American Anthropological Association, archaeologist Clarence W. Weiant revealed that
he would not have made his famous Tres Zapotes discovery, universally considered to be one of the most important Middle American
archaeological finds ever made, were it not for the assistance of a
psychic.[6]

Stephan A. Schwartz, a former editorial staff member of *National
Geographic* magazine and a member of MIT's Secretary of Defense
Discussion Group on Innovation, Technology, and Society, believes
that retrocognition is not only real, but will eventually precipitate a
shift in scientific reality as profound as the shifts that followed the
discoveries of Copernicus and Darwin. Schwartz feels so strongly
about the subject that he has written a comprehensive history of the
partnership between clairvoyants and archaeologists entitled *The Secret Vaults of Time.* "For three-quarters of a century psychic archaeology has been a reality," says Schwartz. "This new approach has
done much to demonstrate that the time and space framework so
crucial to the Grand Material world-view is by no means as absolute
a construct as most scientists believe."[7]

The Past as Hologram

Such abilities suggest that the past is not lost, but still exists in some
form accessible to human perception. Our normal view of the universe
makes no allowance for such a state of affairs, but the holographic
model does. Bohm's notion that the flow of time is the product of a
constant series of unfoldings and enfoldings suggests that as the
present enfolds and becomes part of the past, it does not cease to exist,
but simply returns to the cosmic storehouse of the implicate. Or as
Bohm puts it, "The past is active in the present as a kind of implicate
order."[8]

If, as Bohm suggests, consciousness also has its source in the implicate, this means that the human mind and the holographic record of
the past already exist in the same domain, are, in a manner of speaking, already neighbors. Thus, a shift in the focus of one's attention

may be all that is needed to access the past. Clairvoyants such as McMullen and Ossowiecki may simply be individuals who have an innate knack for making this shift, but again, as with so many of the other extraordinary human abilities we have looked at, the holographic idea suggests that the talent is latent in all of us.

A metaphor for the way the past is stored in the implicate can also be found in the hologram. If each phase of an activity, say a woman blowing a soap bubble, is recorded as a series of successive images in a multiple-image hologram, each image becomes as a frame in a movie. If the hologram is a "white light" hologram—a piece of holographic film whose image can be seen by the naked eye and does not need laser light to become visible—when a viewer walks by the film and changes the angle of his or her perception, he/she will see what amounts to a three-dimensional motion picture of the woman blowing the soap bubble. In other words, as the different images unfold and enfold, they will seem to flow together and present an illusion of movement.

A person who is unfamiliar with holograms might mistakenly assume that the various stages in the blowing of the soap bubble are transitory and once perceived can never be viewed again, but this is not true. The entire activity is always recorded in the hologram, and it is the viewer's changing perspective that provides the illusion that it is unfolding in time. The holographic theory suggests that the same is true of our own past. Instead of fading into oblivion, it too remains recorded in the cosmic hologram and can always be accessed once again.

Another suggestively hologramlike feature of the retrocognitive experience is the three-dimensionality of the scenes that are accessed. For instance, psychic Rich, who can also psychometrize objects, says she knows what Ossowiecki meant when he said that the images he saw were as three-dimensional and real, even more real, than the room in which he was sitting. "It's as if the scene takes over," says Rich. "It's dominant, and once it starts to unfold I actually become a part of it. It's like being in two places at once. I'm aware that I'm sitting in a room, but I'm also in the scene."[9]

Similarly holographic is the nonlocal nature of the ability. Psychics are able to access the past of a particular archaeological site both when they are at the site and when they are many miles removed. In other words, the record of the past does not appear to be stored at any one location, but like the information in a hologram, it is nonlocal and can be accessed from any point in the space-time framework. The

nonlocal aspect of the phenomenon is further underscored by the fact
that some psychics don't even need to resort to psychometry in order
to tune into the past. The famous Kentuckian clairvoyant Edgar Cayce
could tap into the past simply by lying down on a couch in his house
and entering a sleeplike state. He dictated volumes on the history of
the human race and was often startlingly accurate. For example, he
pinpointed the location and described the historical role of the Essene
community at Qumran eleven years before the discovery of the Dead
Sea scrolls (in the caves above Qumran) confirmed his pronounce-
ments.[10]

It is interesting to note that many retrocognitive individuals can also
see the human energy field. When he was a child, Ossowiecki's mother
gave him eye drops in an attempt to get rid of the bands of color he
told her he saw around people, and McMullen also can diagnose a
person's health by looking at their field. This suggests that retrocognition may be linked to the ability to see the subtler and more vibratory
aspects of reality. Put another way, the past may be just one more
thing that is encoded in Pribram's frequency domain, a portion of the
cosmic interference patterns that most of us edit out and only a few
tune into and convert into hologramlike images. "Maybe in the holographic state—in the frequency domain—four thousand years ago is
tomorrow," says Pribram.[11]

Phantoms from the Past

The idea that the past is holographically recorded in the cosmic air-
waves and can occasionally be plucked out by the human mind and
converted into holograms may also explain at least some hauntings.
Many ghostly apparitions appear to be little more than holograms,
three-dimensional recordings of some person or scene from the past.
For example, one theory about ghosts is that they are the soul or spirit
of the deceased individual, but not all ghosts are human. There are
numerous cases on record of individuals seeing phantoms of inanimate
objects as well, a fact that belies the idea that apparitions are discarnate souls. *Phantasms of the Living*, a massive two-volume set of
well-documented reports of hauntings and other paranormal phenomena compiled by the Society for Psychical Research in London, offers
many such examples. For instance, in one case a British military officer

and his family watched as a spectral horse-drawn carriage pulled upon their lawn and stopped. So real was the ghostly carriage that the officer's son walked up to it and saw what appeared to be a female figure inside. The image vanished before he could obtain a better look, and left no horse or wheel tracks.[12]

How common are such experiences? We do not know, but we do know that in the United States and England several studies have shown that from 10 to 17 percent of the general population have seen an apparition, indicating that such phenomena may be far more common than most of us suspect.[13]

The notion that some events leave stronger imprints in the holographic record than others is also supported by the tendency of hauntings to occur at locations where some terrible act of violence or other unusually powerful emotional occurrence has taken place. The literature is filled with apparitions appearing at the sites of murders, military battles, and other kinds of mayhem. This suggests that in addition to images and sounds, the emotions being felt during an event are also recorded in the cosmic hologram. Again it appears that it is the emotional intensity of such events that makes them more prominent in the holographic record, and that allows normal individuals to unwittingly tap into them.

And again, many of these hauntings appear to be less the product of unhappy earthbound spirits, and more just accidental glimpses into the holographic record of the past. This, too, is supported by the literature on the subject. For example, in 1907, and at the prompting of the poet William Butler Yeats, a UCLA anthropologist and religious scholar named W. Y. Evans-Wentz embarked on a two-year journey through Ireland, Scotland, Wales, Cornwall, and Brittany to interview people who had allegedly encountered fairies and other supernatural beings. Evans-Wentz undertook the project because Yeats told him that, as twentieth-century values replaced the old beliefs, encounters with fairies were becoming less frequent and needed to be documented before the tradition was lost completely.

As Evans-Wentz went from village to village interviewing the usually elderly stalwarts of the faith, he discovered that not all of the fairies people encountered in the glens and the moon-dappled meadows were small. Some were tall and looked like normal human beings except they were luminous and translucent and had the curious habit of wearing the clothing of earlier historical periods.

Moreover, these "fairies" often appeared in or around archaeological ruins—burial mounds, standing stones, crumbling sixth-century fortresses, and so on—and participated in activities associated with bygone times. Evans-Wentz interviewed witnesses who had seen fairies that looked like men in Elizabethan dress engaging in hunts, fairies that walked in ghostly processions to and from the remains of old forts, and fairies that rang bells while standing in the ruins of ancient churches. One activity of which the fairies seemed inordinately fond was waging war. In his book *The Fairy-Faith in Celtic Countries* Evans-Wentz presents the testimony of dozens of individuals who claimed to see these spectral conflicts, moonlit meadows thronged with men battling in medieval armor, or desolate fens covered with soldiers in colored uniforms. Sometimes these frays were eerily silent. Sometimes they were full-fledged dins; and, perhaps most haunting of all, sometimes they could only be heard but not seen.

From this, Evans-Wentz concluded that at least some of the phenomena his witnesses were interpreting as fairies were actually some kind of afterimage of events that had taken place in the past. "Nature herself has a memory," he theorized. "There is some indefinable psychic element in the earth's atmosphere upon which all human and physical actions or phenomena are photographed or impressed. Under certain inexplicable conditions, normal persons who are not seers may observe Nature's mental records like pictures cast upon a screen—often like moving pictures."[14]

As for why encounters with fairies were becoming less frequent, a remark made by one of Evans-Wentz's respondents provides a clue. The respondent was an elderly gentleman named John Davies living on the Isle of Man, and after describing numerous sightings of the good people, he stated, "Before education came into the island more people could see the fairies; now very few people can see them."[15] Since "education" no doubt included an anathema against believing in fairies, Davies's remark suggests that it was a change in attitude that caused the widespread retrocognitive abilities of the Manx people to atrophy. Once again this underscores the enormous power our beliefs have in determining which of our extraordinary potentials we manifest and which we do not.

But whether our beliefs allow us to see these hologramlike movies of the past or cause our brains to edit them out, the evidence suggests that they exist nonetheless. Nor are such experiences limited to Celtic countries. There are reports of witnesses seeing phantom soldiers dressed in ancient Hindu costumes in India.[16] In Hawaii, such ghostly displays are well known and books on the islands are filled with accounts of individuals who have seen phantom processions of Hawaiian warriors in feather cloaks marching along with war clubs and torches.[17] Sightings of spectral armies fighting equally phantasmal battles are even mentioned in ancient Assyrian texts.[18]

Occasionally historians are able to recognize the event being replayed. At four in the morning on August 4, 1951, two English women vacationing in the seaside village of Puys, France, were awakened by the sound of gunfire. They raced to the window but were shocked to find that the village and the sea beyond were calm and devoid of any activity that might account for what they were hearing. The British Society for Psychical Research investigated and discovered that the women's chronology of events mirrored exactly military records of a raid the Allies had made against the Germans at Puys on August 19, 1942. The women, it seemed, had heard the sound of a slaughter that had taken place nine years earlier.[19]

Although the dark intensity of such events gives them a higher profile in the holographic landscape, we must not forget that contained within the shimmering holographic record of the past are all the joys of the human race as well. It is, in essence, a library of all that ever was, and learning to tap into this dazzling and infinite treasure-trove on a more massive and systematic scale could expand our knowledge of both ourselves and the universe in ways we have not yet dared dream. The day may come when we can manipulate reality like the crystal in Bohm's analogy, causing what is real and what is invisible to shift kaleidoscopically and calling up images of the past with the same ease that we now call up a program on our computer. But even this is not all that a more holographic understanding of time may offer.

The Holographic Future

As disconcerting as having access to the entire past is, it pales beside the notion that the future is also accessible in the cosmic hologram. Still, there is an enormous body of evidence that proves at least some future events are as easy to see as past events.

This has been amply demonstrated in literally hundreds of studies. In the 1930s J. B. and Louisa Rhine discovered that volunteers could guess what cards would be drawn randomly from a deck with a success rate that was better than chance by odds of three million to one.[20] In the 1970s Helmut Schmidt, a physicist at Boeing Aircraft in Seattle, Washington, invented a device that enabled him to test whether people could predict random subatomic events. In repeated tests with three volunteers and over sixty thousand trials, he obtained results that were one billion to one against chance.[21]

In his work at the Dream Laboratory at Maimonides Medical Center, Montague Ullman, along with psychologist Stanley Krippner and researcher Charles Honorton, produced compelling evidence that accurate precognitive information can also be obtained in dreams. In their study, volunteers were asked to spend eight consecutive nights at the sleep laboratory, and each night they were asked to try to dream about a picture that would be chosen at random the next day and shown to them. Ullman and his colleagues hoped to get one success out of eight, but found that some subjects could score as many as five "hits" out of eight.

For example, after waking, one volunteer said that he had dreamed of "a large concrete building" from which a "patient" was trying to escape. The patient had a white coat on like a doctor's coat and had gotten only "as far as the archway." The painting chosen at random the next day turned out to be Van Gogh's *Hospital Corridor at St. Rémy,* a watercolor depicting a lone patient standing at the end of a bleak and massive hallway and quickly exiting through a door beneath an archway.[22]

In their remote-viewing experiments at Stanford Research Institute, Puthoff and Targ found that, in addition to being able to psychically describe remote locations that experimenters were visiting in the present, test subjects could also describe locations experimenters would be visiting in the future, *before* the locations had even been decided upon. In one instance, for example, an unusually talented subject named Hella Hammid, a photographer by vocation, was asked to describe the spot Puthoff would be visiting one-half hour hence. She concentrated and said she could see him entering "a black iron triangle." The triangle was "bigger than a man," and although she did not know precisely what it was, she could hear a rhythmic squeaking sound occurring "about once a second."

Ten minutes before she did this, Puthoff had set out on a half-hour drive in the Menlo Park and Palo Alto areas. At the end of the half hour, and well after Hammid had recorded her perception of the black iron triangle, Puthoff took out ten sealed envelopes containing ten

different target locations. Using a random number generator, he chose
one at random. Inside was the address of a small park about six miles
from the laboratory. He drove to the park, and when he got there he
found a children's swing—the black iron triangle—and walked into its
midst. When he sat down in the swing it squeaked rhythmically as it
swung back and forth.[23]

Puthoff and Targ's precognitive remote-viewing findings have been
duplicated by numerous laboratories around the world, including Jahn
and Dunne's research facility at Princeton. Indeed, in 334 formal trials
Jahn and Dunne found that volunteers were able to come up with
accurate precognitive information 62 percent of the time.[24]

Even more dramatic are the results of the so-called "chair tests," a
famous series of experiments devised by Croiset. First, the experimenter would randomly select a chair from the seating plan for an
upcoming public event in a large hall or auditorium. The hall could be
located in any city in the world and only events that did not have
reserved seating qualified. Then, without telling Croiset the name or
location of the hall, or the nature of the event, the experimenter would
ask the Dutch psychic to describe who would be sitting in the seat
during the evening in question.

Over the course of a twenty-five-year period, numerous investigators in both Europe and America put Croiset through the rigors of the
chair test and found that he was almost always capable of giving an
accurate and detailed description of the person who would be sitting
in the chair, including describing their gender, facial features, dress,
occupation, and even incidents from their past.

For instance, on January 6, 1969, in a study conducted by Dr. Jule
Eisenbud, a clinical professor of psychiatry at the University of Colorado Medical School, Croiset was told that a chair had been chosen for
an event that would take place on January 23, 1969. Croiset, who was
in Utrecht, Holland, at the time, told Eisenbud that the person who
would sit in the chair would be a man five feet nine inches in height
who brushed his black hair straight back, had a gold tooth in his lower
jaw, a scar on his big toe, who worked in both science and industry,
and sometimes got his lab coat stained by a greenish chemical. On
January 23, 1969, the man who sat down in the chair, which was in an
auditorium in Denver, Colorado, fit Croiset's description in every way
but one. He was not five feet nine, but five feet nine and three-quarters.[25]

The list goes on and on.

What is the explanation for such findings? Krippner believes that Bohm's assertion that the mind can access the implicate order is one explanation.[26] Both Puthoff and Targ feel that nonlocal quantum interconnectedness plays a role in precognition, and Targ has asserted that during a remote-viewing experience the mind appears to be able to access some kind of "holographic soup," or domain, in which all points are infinitely interconnected not only in space, but in time as well.[27]

Dr. David Loye, a clinical psychologist and a former member of the Princeton and UCLA medical school faculties, agrees. "For those pondering the puzzle of precognition, the Pribram-Bohm holographic mind theory seems to offer the greatest hope yet for progress toward the sought-for solution," he states. Loye, who is currently codirector of the Institute for Future Forecasting in northern California, knows whereof he speaks. He has spent the last two decades investigating precognition and the art of forecasting in general, and develops techniques to enable people to get in touch with their own intuitive awareness of the future.[28]

The hologramlike nature of many precognitive experiences provides further evidence that the ability to foresee the future is a holographic phenomenon. As with retrocognition, psychics report that precognitive information often appears to them in the form of three-dimensional images. Cuban-born psychic Tony Cordero says that when he sees the future it's like watching a movie in his mind. Cordero saw one of the first such movies when he was a child and had a vision of the Communist takeover of Cuba. "I told my family that I saw red flags all over Cuba and they were going to have to leave the country and that a lot of members of the family were going to be shot," says Cordero. "I actually saw relatives being shot. I could smell smoke and hear the sound of gunfire. I feel like I'm in the situation. I can hear people talking but they cannot hear or see me. It's like traveling into time or something."[29]

The words psychics use to describe their experiences are also similar to Bohm's. Garrett described clairvoyance as "an intensely acute sensing of some aspects of life in operation, and since at clairvoyant levels time is *undivided and whole* [italics added], one often perceives the object or event in its past, present and/or future phases in abruptly swift successions."[30]

We Are All Precognitive

Bohm's assertion that every human consciousness has its source in the implicate implies that we all possess the ability to access the future, and this is also supported by the evidence. Jahn and Dunne's discovery that even normal individuals do well in precognitive remote-viewing tests is one indication of the widespread nature of the ability. Numerous other findings, both experimental and anecdotal, provide additional evidence. In a 1934 BBC broadcast Dame Edith Lyttelton, a member of the politically and socially prominent Balfour family in England and the president of the British Society for Psychical Research, invited listeners to send in accounts of their own precognitive experiences. She was inundated with mail, and even after eliminating the cases that did not have corroborative evidence, she still had enough to fill a volume on the subject.[31] Similarly, surveys conducted by Louisa Rhine revealed that precognitions occur more frequently than any other kind of psychic experience.[32]

Studies also show that precognitive visions tend to be of tragedies, with premonitions of unhappy events outnumbering happy ones by a ratio of four to one. Presentiments of death predominate, with accidents coming in second, and illnesses third.[33] The reason for this seems obvious. We are so thoroughly conditioned to believe that perceiving the future is *not* possible, our natural precognitive abilities have gone dormant. Like the superhuman strengths individuals display during life-threatening emergencies, they only spill over into our conscious minds during times of crisis—when someone near to us is about to die; when our children or some other loved one is in danger, and so on. That our "sophisticated" understanding of reality is responsible for our inability to both grasp and utilize the true nature of our relationship with time is evident in the fact that primitive cultures nearly always score better on ESP tests than so-called civilized cultures.[34]

Further evidence that we have relegated our innate precognitive abilities to the hinterlands of the unconscious can be found in the close association between premonitions and dreams. Studies show that from 60 to 68 percent of all precognitions occur during dreaming.[35] We may have banished our ability to see the future from our conscious minds, but it is still very active in the deeper strata of our psyches.

Tribal cultures are well aware of this fact, and shamanic traditions almost universally stress how important dreaming is in divining the future. Even our most ancient writings pay homage to the premonitory power of dreams, as is evidenced in the biblical account of Pharaoh's dream of seven fat and seven lean cows. The antiquity of such traditions indicates that the tendency of premonitions to occur in dreams is due to more than just our current skeptical attitude toward precognition. The proximity the unconscious mind has to the atemporal realm of the implicate may also play a role. Because our dreaming self is deeper in the psyche than our conscious self—and thus closer to the primal ocean in which past, present, and future become one—it may be easier for it to access information about the future.

Whatever the reason, it should come as no surprise that other methods for accessing the unconscious can also produce precognitive information. For example, in the 1960s Karlis Osis and hypnotist J. Fahler found that hypnotized subjects scored significantly higher on precognition tests than nonhypnotized subjects.[36] Other studies have also confirmed the ESP-enhancing effects of hypnosis.[37] However, no amount of dry statistical data has the impact of an example from real life. In his book *The Future Is Now: The Significance of Precognition*, Arthur Osborn records the results of a hypnosis-precognition experiment involving the French actress Irene Muza. After being hypnotized and asked if she could see her future, Muza replied, "My career will be short: I dare not say what my end will be: it will be terrible."

Startled, the experimenters decided not to tell Muza what she had reported and gave her a posthypnotic suggestion to forget everything she had said. When she awakened from her trance she had no memory of what she had predicted for herself. Even if she had known, it would not have caused the type of death she suffered. A few months later her hairdresser accidentally spilled some mineral spirits on a lighted stove, causing Muza's hair and clothing to be set on fire. Within seconds she was engulfed in flames and died in a hospital a few hours later.[38]

Hololeaps of Faith

The events that befell Irene Muza raise an important question. If Muza had known about the fate she had predicted for herself, would she have been able to avoid it? Put another way, is the future frozen and completely predetermined, or can it be changed? At first blush, the existence of precognitive phenomena seems to indicate that the former is the case, but this would be a very disturbing state of affairs. If the future is a hologram whose every detail is already fixed, it means that we have no free will. We are all just puppets of destiny moving mindlessly through a script that has already been written.

Fortunately the evidence overwhelmingly indicates that this is not the case. The literature is filled with examples of people who were able to use their precognitive glimpses of the future to avoid disasters, instances in which individuals correctly foresaw the crash of a plane and avoided death by not getting on, or had a vision of their children being drowned in a flood and moved them out of harm's way just in the nick of time. There are nineteen documented cases of people who had precognitive glimpses of the sinking of the *Titanic*—some were experienced by passengers who paid attention to their premonitions and survived, some were experienced by passengers who ignored their forebodings and drowned, and some were experienced by individuals who were not in either of these two categories.[39]

Such incidents strongly suggest that the future is not set, but is plastic and can be changed. But this view also brings with it a problem. If the future is still in a state of flux, what is Croiset tapping into when he describes the individual who will sit down in a particular chair seventeen days hence? How can the future both exist and not exist?

Loye provides a possible answer. He believes that reality *is* a giant hologram, and in it the past, present, and future are indeed fixed, at least up to a point. The rub is that it is not the only hologram. There are many such holographic entities floating in the timeless and spaceless waters of the implicate, jostling and swimming around one another like so many amoebas. "Such holographic entities could also be visualized as parallel worlds, parallel universes," says Loye.

Thus, the future of any given holographic universe *is* predetermined, and when a person has a precognitive glimpse of the future, they are tuning into the future of that particular hologram only. But like amoebas, these holograms also occasionally swallow and engulf each other, melding and bifurcating like the protoplasmic globs of energy that they really are. Sometimes these jostlings jolt us and are responsible for the premonitions that from time to time engulf us. And when we act upon a premonition and appear to alter the future, what we are really doing is leaping from one hologram to another. Loye calls these *intra* holographic leaps "hololeaps" and feels that they are what provides us with our true capacity for both insight and freedom.[40]

Bohm sums up the same situation in a slightly different manner. "When people dream of accidents correctly and do not take the plane or ship, it is not the actual future that they were seeing. It was merely something in the present which is implicate and moving toward making that future. In fact, the future they saw differed from the actual future because they altered it. Therefore I think it's more plausible to say that, if these phenomena exist, there's an anticipation of the future in the implicate order in the present. As they used to say, coming events cast their shadows in the present. Their shadows are being cast deep in the implicate order."[41]

Bohm's and Loye's descriptions seem to be two different ways of trying to express the same thing—a view of the future as a hologram that is substantive enough for us to perceive it, but malleable enough to be susceptible to change. Others have used still different words to sum up what appears to be the same basic thought. Cordero describes the future as a hurricane that is beginning to form and gather momentum, becoming more concrete and unavoidable as it approaches.[42] Ingo Swann, a gifted psychic who has produced impressive results in various studies, including Puthoff and Targ's remote-viewing research, speaks of the future as composed of "crystallizing possibilities."[43] The Hawaiian kahunas, widely esteemed for their precognitive powers, also speak of the future as fluid, but in the process of "crystallizing," and believe that great world events are crystallized furthest in advance, as are the most important events in a person's life, such as marriage, accidents, and death.[44]

The numerous premonitions that are now known to have preceded both the Kennedy assassination and the Civil War (even George Washington had a precognitive vision of a future civil war somehow involving "Africa," the issue that all men are "brethren," and the word *Union*[45]) seem to corroborate this kahuna belief.

Loye's notion that there are many separate holographic futures and we choose which events are going to manifest and which are not by leaping from one hologram to another carries with it another implication. Choosing one holographic future over another is essentially the same as creating the future. As we have seen, there is a good deal of evidence suggesting that consciousness plays a significant role in creating the here and now. But if the mind can stray beyond the boundaries of the present and occasionally stalk the misty landscape of the future, do we have a hand in creating future events as well? Put another way, are the vagaries of life truly random, or do we play a role in literally sculpting our own destiny? Remarkably, there is some intriguing evidence that the latter may be the case.

The Shadowy Stuff of the Soul

Dr. Joel Whitton, a professor of psychiatry at the University of Toronto Medical School, has also used hypnosis to study what people unconsciously know about themselves. However, instead of asking them about their future, Whitton, who is an expert in clinical hypnosis and also holds a degree in neurobiology, asks them about their past, their distant past to be exact. For the last several decades Whitton has quietly and without fanfare been gathering evidence suggestive of reincarnation.

Reincarnation is a difficult subject, for so much silliness has been presented about it that many people dismiss it out of hand. Most do not realize that in addition to (and one might even say in spite of) the sensational claims of celebrities and the stories of reincarnated Cleopatras that garner most of the media attention, there is a good deal of serious research being done on reincarnation. In the last several decades a small but growing number of highly credentialed researchers has compiled an impressive body of evidence on the subject. Whitton is one of these researchers.

The evidence does not prove that reincarnation exists, nor is it the intention of this book to make such an argument. In fact, it is difficult to imagine what might constitute perfect proof of reincarnation. Rather, the findings that will be touched upon here are offered only as intriguing possibilities and because they are relevant to our current discussion. Thus, they deserve our open-minded consideration.

The main thrust of Whitton's hypnosis research is based on a simple and startling fact. When individuals are hypnotized, they often remember what appear to be memories of previous existences. Studies have shown that over 90 percent of all hypnotizable individuals are able to recall these apparent memories.[46] The phenomenon is widely recognized, even by skeptics. For example, the psychiatry textbook *Trauma, Trance and Transformation* warns fledgling hypnotherapists not to be surprised if such memories surface spontaneously in their hypnotized patients. The author of the text rejects the idea of rebirth but does note that such memories can have remarkable healing potential nonetheless.[47]

The meaning of this phenomenon is, of course, hotly debated. Many researchers argue that such memories are fantasies or fabrications of the unconscious mind, and there is no doubt that this is sometimes the case, especially if the hypnotic session or "regression" is conducted by an unskilled hypnotist who does not know the proper questioning techniques required to safeguard against eliciting fantasies. But there are also numerous cases on record in which individuals have, under the guidance of skilled professionals, produced memories that do not appear to be fantasies. The evidence assembled by Whitton falls into this category.

To conduct his research, Whitton gathered together a core group of roughly thirty people. These included individuals from all walks of life, from truck drivers to computer scientists, some of whom believed in reincarnation and some of whom did not. He then hypnotized them individually and spent literally thousands of hours recording everything they had to say about their alleged previous existences.

Even in its broad strokes the information was fascinating. One striking aspect was the degree of agreement between the subjects' experiences. All reported numerous past lives, some as many as twenty to twenty-five, although a practical limit was reached when Whitton regressed them to what he calls their "caveman existences," when one lifetime became indistinguishable from the next.[48] All reported that gender was not specific to the soul, and many had lived at least one life as the opposite sex. And all reported that the purpose of life was to evolve and learn, and that multiple existences facilitated this process.

Whitton also found evidence that strongly suggested the experiences were actual past lives. One unusual feature was the ability the memories had to explain a wide range of seemingly unrelated events and experiences in the subjects' current lives. For example, one man, a psychologist born and raised in Canada, had possessed an inexplicable British accent as a child. He also had an irrational fear of breaking his leg, a phobia of air travel, a terrible nail-biting problem, an obsessive fascination with torture, and as a teenager had had a brief and enigmatic vision of being in a room with a Nazi officer, shortly after operating the pedals of a car during a driving test. Under hypnosis the man recalled being a British pilot during World War II. While on a mission over Germany his plane was hit by a shower of bullets, one of which penetrated the fuselage and broke his leg. This in turn caused him to lose control of the plane's foot pedals, forcing him to crash-land. He was subsequently captured by the Nazis, tortured for information by having his nails pulled out, and died a short time later.[49]

Many of the subjects also experienced profound psychological and physical healings as a result of the traumatic past-life memories they unearthed, and gave uncannily accurate historical details about the times in which they had lived. Some even spoke languages unknown to them. While reliving an apparent past life as a Viking, one man, a thirty-seven-year-old behavioral scientist, shouted words that linguistic authorities later identified as Old Norse.[50] After being regressed to an ancient Persian lifetime, the same man began to write in a spidery, Arabic-style script that an expert in Near Eastern languages identified as an authentic representation of Sassanid Pahlavi, a long-extinct Mesopotamian tongue that flourished between A.D. 226 and 651.[51]

But Whitton's most remarkable discovery came when he regressed subjects to the interim between lives, a dazzling, light-filled realm in which there was "no such thing as time or space as we know it."[52] According to his subjects, part of the purpose of this realm was to allow them *to plan their next life, to literally sketch out the important events and circumstances that would befall them in the future.* But this process was not simply some fairy-tale exercise in wish fulfillment. Whitton found that when individuals were in the between-life realm, they entered an unusual state of consciousness in which they were acutely self-aware and had a heightened moral and ethical sense. In addition, they no longer possessed the ability to rationalize away any of their faults and misdeeds, and saw themselves with total honesty. To distinguish it from our normal everyday consciousness, Whitton calls this intensely conscientious state of mind "metaconsciousness."

Thus, when subjects planned their next life, they did so with a sense of moral obligation. They would choose to be reborn with people whom they had wronged in a previous life so they would have the opportunity to make amends for their actions. They planned pleasant encounters with "soul mates," individuals with whom they had built a loving and mutually beneficial relationship over many lifetimes; and they scheduled "accidental" events to fulfill still other lessons and purposes. One man said that as he planned his next life he visualized "a sort of clockwork instrument into which you could insert certain parts in order for specific consequences to follow."[53]

These consequences were not always pleasant. After being regressed to a metaconscious state, a woman who had been raped when she was thirty-seven revealed that she had actually planned the event before she had come into this incarnation. As she explained, it had been necessary for her to experience a tragedy at that age in order to force her to change her "entire soul complexion" and thus break through to a deeper and more positive understanding of the meaning of life.[54] Another subject, a man afflicted with a serious and life-threatening kidney disease, disclosed that he had chosen the illness to punish himself for a past-life transgression. However, he also revealed that dying from the kidney disease was not part of his script, and before he had come into this life he had also arranged to encounter someone or something that would help him remember this fact and hence enable him to heal both his guilt and his body. True to his word, after he started his sessions with Whitton he experienced a near-miraculous complete recovery.[55]

Not all of Whitton's subjects were so eager to learn about the future their metaconscious selves had laid out for them. Several censored their own memories and asked Whitton to please give them posthypnotic instructions *not* to remember anything that they had said during trance. As they explained, they did not want to be tempted to tamper with the script their metaconscious selves had written for them.[56]

This is an astounding idea. Is it possible that our unconscious mind is not only aware of the rough outline of our destiny, but actually steers us toward its fulfillment? Whitton's research is not the only evidence that this may be the case. In a statistical study of 28 serious U.S. railroad accidents, parapsychologist William Cox found that significantly fewer people took trains on accident days than on the same day in previous weeks.[57]

Cox's finding suggests that we all may be constantly unconsciously precognizing the future and making decisions based on that information: some of us opting to avoid mishap, and perhaps some—like the woman who chose to experience a personal tragedy and the man who elected to endure a kidney disease—choosing to experience negative situations to fulfill other unconscious designs and purposes. "Carefully or haphazardly, we choose our earthly circumstances," says Whitton. "The message of metaconsciousness is that the life situation of every human being is neither random nor inappropriate. Seen objectively from the interlife, every human experience is simply another lesson in the cosmic classroom."[58]

It is important to note that the existence of such unconscious agendas does not mean that our lives are rigidly predestined and all fates unavoidable. The fact that many of Whitton's subjects asked not to remember what they said under hypnosis implies again that the future is only roughly outlined and still subject to change.

Whitton is not the only reincarnation researcher who has uncovered evidence that our unconscious has more of a hand in our lives than we may realize. Another is Dr. Ian Stevenson, a professor of psychiatry at the University of Virginia Medical School. Instead of using hypnosis, Stevenson interviews young children who have spontaneously remembered apparent previous existences. He has spent more than thirty years in this pursuit and has collected and analyzed thousands of cases from all over the globe.

According to Stevenson, spontaneous past-life recall is relatively common among children, so common that the number of cases that seem worth considering far exceeds his staff's ability to investigate them. Generally children are between the ages of two and four when they start talking about their "other life," and frequently they remember dozens of particulars, including their name, the names of family members and friends, where they lived, what their house looked like, what they did for a living, how they died, and even obscure information such as where they hid money before they died and, in cases involving murder, sometimes even who killed them.[59]

Indeed, frequently their memories are so detailed Stevenson is able to track down the identity of their previous personality and verify virtually everything they have said. He has even taken children to the area in which their past incarnation lived, and watched as they navigated effortlessly through strange neighborhoods and correctly identified their former house, belongings, and past-life relatives and friends.

Like Whitton, Stevenson has gathered an enormous amount of data suggestive of reincarnation, and to date has published six volumes on his findings.[60] And like Whitton, he also has found evidence that the unconscious plays a far greater role in our makeup and destiny than we have hitherto suspected.

He has corroborated Whitton's finding that we are frequently reborn with individuals we have known in previous existences, and that the guiding force behind our choices is often affection or a sense of guilt or indebtedness.[61] He agrees that personal responsibility, not chance, is the arbiter of our fate. He has found that although a person's material conditions can vary greatly from one life to the next, their moral conduct, interests, aptitudes, and attitudes remain the same. Individuals who were criminals in their previous existence tend to be drawn to criminal behavior again; people who were generous and kind continue to be generous and kind, and so on. From this Stevenson concludes that it is not the outward trappings of life that matter, but the inner ones, the joys, sorrows, and "inner growths" of the personality, that appear to be most important.

Most significant of all, he found no compelling evidence of "retributive karma," or any indication that we are cosmically punished for our sins. "There is then—if we judge by the evidence of the cases—no external judge of our conduct and no being who shifts us from life to life according to our deserts. If this world is (in Keats's phrase) 'a vale of soul-making,' we are the makers of our own souls," states Stevenson.[62]

Stevenson has also uncovered a phenomenon that did not turn up in Whitton's study, a discovery that provides even more dramatic evidence of the power the unconscious mind has to sculpt and influence our life circumstances. He has found that a person's previous incarnation can apparently affect the very shape and structure of their current physical body. He has discovered, for example, that Burmese children who remember previous lives as British or American Air Force pilots shot down over Burma during World War II all have fairer hair and complexions than their siblings.[63]

He has also found instances in which distinctive facial features, foot deformities, and other characteristics have carried over from one life to the next.[64] Most numerous among these are physical injuries carrying over as scars or birthmarks. In one case, a boy who remembered being murdered in his former life by having his throat slit still had a long reddish mark resembling a scar across his neck.[65] In another, a boy who remembered committing suicide by shooting himself in the head in his past incarnation still had two scarlike birthmarks that lined up perfectly along the bullet's trajectory, one where the bullet had entered and one where it had exited.[66] And in another, a boy had a birthmark resembling a surgical scar complete with a line of red marks resembling stitch wounds, in the exact location where his previous personality had had surgery.[67]

In fact, Stevenson has gathered hundreds of such cases and is currently compiling a four-volume study of the phenomenon. In some of the cases he has even been able to obtain hospital and/or autopsy reports of the deceased personality and show that such injuries not only occurred, but were in the exact location of the present birthmark or deformity. He feels that such marks not only provide some of the strongest evidence in favor of reincarnation, but also suggest the existence of some kind of intermediate nonphysical body that functions as a carrier of these attributes between one life and the next. He states, "It seems to me that the imprint of wounds on the previous personality must be carried between lives on some kind of an extended body which in turn acts as a template for the production on a new physical body of birthmarks and deformities that correspond to the wounds on the body of the previous personality."[68]

Stevenson's theorized "template body" echoes Tiller's assertion that the human energy field is a holographic template that guides the form and structure of the physical body. Put another way, it is a kind of three-dimensional blueprint around which the physical body forms. Similarly, his findings regarding birthmarks add further support to the idea that we are at heart just images, holographic constructs, created by thought.

Stevenson has also noted that although his research suggests that we are the creators of our own lives and, to a certain extent, our own bodies, our participation in this process is so passive as to be almost involuntary. Deep strata of the psyche appear to be involved in these choices, strata that are much more in touch with the implicate. Or as Stevenson puts it, "Levels of mental activity far deeper than those that regulate the digestion of our supper in our stomach [and] our ordinary breathing must govern these processes."[69]

As unorthodox as many of Stevenson's conclusions are, his reputation as a careful and thorough investigator has gained him respect in some unlikely quarters. His findings have been published in such distinguished scientific periodicals as the *American Journal of Psychiatry*, the *Journal of Nervous and Mental Disease*, and the *International Journal of Comparative Sociology*. And in a review of one of his works the prestigious *Journal of the American Medical Association* stated that he has "painstakingly and unemotionally collected a detailed series of cases in which the evidence for reincarnation is difficult to understand on any other grounds. . . . He has placed on record a large amount of data that cannot be ignored."[70]

Thought as Builder

As with so many of the "discoveries" we have looked at, the idea that
some deeply unconscious and even spiritual part of us can reach across
the boundaries of time and is responsible for our destiny can also be
found in many shamanic traditions and other sources. According to the
Batak people of Indonesia, everything a person experiences is determined by his or her soul, or *tondi,* which reincarnates from one body
to the next and is a medium capable of reproducing not only the
behavior, but the physical attributes of the person's former self.[71] The
Ojibway Indians also believed a person's life is scripted by an invisible
spirit or soul and is laid out in a manner that promotes growth and
development. If a person dies without completing all the lessons they
need to learn, their spirit body returns and is reborn in another physical body.[72]

The kahunas call this invisible aspect the *aumakua,* or "high self."
Like Whitton's metaconsciousness, it is the unconscious portion of a
person that can see the parts of the future that are crystallized, or
"set." It is also the part of us that is responsible for creating our
destiny, but it is not alone in this process. Like many of the researchers
mentioned in this book, the kahunas believed that thoughts are things
and are composed of a subtle energetic substance they called *kino
mea,* or "shadowy body stuff." Hence, our hopes, fears, plans, worries, guilts, dreams, and imaginings do not vanish after leaving our
mind, but are turned into thought forms, and these, too, become some
of the rough strands from which the high self weaves our future.

Most people are not in charge of their own thoughts, said the kahunas, and constantly bombard their high self with an uncontrolled and
contradictory mixture of plans, wishes, and fears. This confuses the
high self and is why most people's lives appear to be equally haphazard and uncontrolled. Powerful kahunas who were in open communication with their high selves were said to be able to help a person remake
his or her future. Similarly, it was considered extremely important
that people take time out at frequent intervals to think about their
lives and visualize in concrete terms what they wished to happen to
themselves. By doing this the kahunas asserted that people can more
consciously control the events that befall them and make their own
future.[73]

In an idea that is reminiscent of Tiller and Stevenson's notion of a subtle intermediary body, the kahunas believed this shadowy body stuff also forms a template upon which the physical body is molded. Again it was said that kahunas who were in extraordinary attunement with their high self could sculpt and reform the shadowy body stuff, and hence the physical body, of another person and this was how miraculous healings were effected.[74] This view also provides an interesting parallel to some of our own conclusions as to why thoughts and images have such a powerful impact on health.

The tantric mystics of Tibet referred to the "stuff" of thoughts as *tsal* and held that every mental action produced waves of this mysterious energy. They believed the entire universe is a product of the mind and is created and animated by the collective *tsal* of all beings. Most people are unaware that they possess this power, said the Tantrists, because the average human mind functions "like a small puddle isolated from the great ocean." Only great yogis skilled at contacting the deeper levels of the mind were said to be able consciously to utilize such forces, and one of the things they did to achieve this goal was to visualize repeatedly the desired creation. Tibetan tantric texts are filled with visualization exercises, or "sadhanas," designed for such purposes, and monks of some sects, such as the Kargyupa, would spend as long as seven years in complete solitude, in a cave or a sealed room, perfecting their visualization abilities.[75]

The twelfth-century Persian Sufis also stressed the importance of visualization in altering and reshaping one's destiny, and called the subtle matter of thought *alam almithal*. Like many clairvoyants, they believed that human beings possess a subtle body controlled by chakralike energy centers. They also held that reality is divided into a series of subtler planes of being, or *Hadarat*, and that the plane of being directly adjacent to this one was a kind of template reality in which the *alam almithal* of one's thoughts formed into idea-images, which in turn eventually determined the course of one's life. The Sufis also added a twist of their own. They felt the heart chakra, or *himma*, was the agent responsible for this process, and that control of the heart chakra was therefore a prerequisite for controlling one's destiny.[76]

Edgar Cayce also spoke of thoughts as tangible things, a finer form of matter and, when he was in trance, repeatedly told his clients that their thoughts created their destiny and that "thought is the builder." In his view, the thinking process is like a spider constantly spinning,

constantly adding to its web. Every moment of our lives we are creating the images and patterns that give our future energy and shape,
said Cayce.[77]

Paramahansa Yogananda advised people to visualize the future
they desired for themselves and charge it with the "energy of concentration." As he put it, "Proper visualization by the exercise of concentration and willpower enables us to materialize thoughts, not only as
dreams or visions in the mental realm, but also as experiences in the
material realm."[78]

Indeed, such ideas can be found in a wide range of disparate
sources. "We are what we think," said the Buddha. "All that we are
arises with our thoughts. With our thoughts we make the world."[79]
"As a man acts, so does he become. As a man's desire is, so is his
destiny," states the Hindu pre-Christian Brihadaranyaka Upanishad.[80] "All things in the world of Nature are not controlled by Fate
for the soul has a principle of its own," said the fourth-century Greek
philosopher Iamblichus.[81] "Ask and it will be given you. . . . If ye have
faith, nothing shall be impossible unto you," states the Bible.[82] And,
"The destiny of a person is connected with those things he himself
creates and does," wrote Rabbi Steinsaltz in the kabbalistic *Thirteen-
Petaled Rose.*[83]

An Indication of Something Deeper

Even today the idea that our thoughts create our destiny is still very
much in the air. It is the subject of best-selling self-help books such
as Shakti Gawain's *Creative Visualization* and Louise L. Hay's *You
Can Heal Your Life.* Hay, who says she cured herself of cancer by
changing her mental patterning, gives hugely successful workshops
on her techniques. It is the main philosophy inherent in many popular
"channeled" works such as *A Course in Miracles* and Jane Roberts's
Seth books.

It is also being embraced by some eminent psychologists. Jean
Houston, a past president of the Association for Humanistic Psychol-
ogy and current Director of the Foundation for Mind Research in
Pomona, New York, discusses the idea at length in her book *The
Possible Human.* Houston also gives a variety of visualization exer-
cises in the work and even calls one "Orchestrating the Brain and
Entering the Holoverse."[84]

Another book that draws heavily on the holographic model to support the idea that we can use visualization to reshape our future is Mary Orser and Richard A. Zarro's *Changing Your Destiny*. In addition, Zarro is the founder of Futureshaping Technologies, a company that gives seminars on "futureshaping" techniques to businesses, and numbers both Panasonic and the International Banking and Credit Association among its clients.[85]

Former astronaut Edgar Mitchell, the sixth man to walk on the moon and a longtime explorer of inner as well as outer space, has taken a similar tack. In 1973 he founded the Institute of Noetic Sciences, a California-based organization devoted to researching such powers of the mind. The institute is still going strong, and current projects include a massive study of the mind's role in miraculous healings and spontaneous remissions, and a study of the role consciousness plays in creating a positive global future. "We create our own reality because our inner emotional—our subconscious—reality draws us into those situations from which we learn," states Mitchell. "We experience it as strange things happening to us [and] we meet the people in our lives that we need to learn from. And so we create these circumstances at a very deep metaphysical and subconscious level."[86]

Is the current popularity of the idea that we create our own destiny just a fad, or is its presence in so many different cultures and times an indication of something much deeper, a sign that it is something all human beings intuitively know is true? At present this question remains unanswered, but in a holographic universe—a universe in which the mind *participates* with reality and in which the innermost stuff of our psyches can register as synchronicities in the objective world—the notion that we are also the sculptors of our own fate is not so far-fetched. It even seems probable.

Three Last Pieces of Evidence

Before concluding, three last pieces of evidence deserve to be looked at. Although not conclusive, each offers a peek at still other time-transcending abilities consciousness may possess in a holographic universe.

MASS DREAMS OF THE FUTURE

Another past-life researcher who turned up evidence suggestive that the mind has a hand in creating one's destiny was the late San Francisco–based psychologist Dr. Helen Wambach. Wambach's approach was to hypnotize groups of people in small workshops, regress them to specified time periods, and ask them a predetermined list of questions about their sex, clothing style, occupation, utensils used in eating, and so on. Over the course of her twenty-nine-year investigation of the past-life phenomenon, she hypnotized literally thousands of individuals and amassed some impressive findings.

One criticism leveled against reincarnation is that people only seem to remember past lives as famous or historical personages. Wambach, however, found that more than 90 percent of her subjects recalled past lives as peasants, laborers, farmers, and primitive food gatherers. Less than 10 percent remembered incarnations as aristocrats, and none remembered being anyone famous, a finding that argues against the notion that past-life memories are fantasies.[87] Her subjects were also extraordinarily accurate when it came to historical details, even obscure ones. For instance, when people remembered lives in the 1700s, they described using a three-pronged fork to eat their evening meals, but after 1790 they described most forks as having four prongs, an observation that correctly reflects the historical evolution of the fork. Subjects were equally accurate when it came to describing clothing and footwear, types of foods eaten, et cetera.[88]

Wambach discovered she could also *progress* people to future lives. Indeed, her subjects' descriptions of coming centuries were so fascinating she conducted a major future-life-progression project in France and the United States. Unfortunately, she passed away before completing the study, but psychologist Chet Snow, a former colleague of Wambach's, carried on her work and recently published the results in a book entitled *Mass Dreams of the Future*.

When the reports of the 2,500 people who participated in the project were tallied, several interesting features emerged. First, virtually all of the respondents agreed that the population of the earth had decreased dramatically. Many did not even find themselves in physical bodies in the various future time periods specified, and those who did noted that the population was much smaller than it is today.

In addition, the respondents divided up neatly into four categories, each relating a different future. One group described a joyless and sterile future in which most people lived in space stations, wore silvery suits, and ate synthetic food. Another, the "New Agers," reported living happier and more natural lives in natural settings, in harmony with one another, and in dedication to learning and spiritual development. Type 3, the "hi-tech urbanites," described a bleak mechanical future in which people lived in underground cities and cities enclosed in domes and bubbles. Type 4 described themselves as post-disaster survivors living in a world that had been ravaged by some global, possibly nuclear, disaster. People in this group lived in homes ranging from urban ruins to caves to isolated farms, wore plain handsewn clothing that was often made of fur, and obtained much of their food by hunting.

What is the explanation? Snow turns to the holographic model for the answer, and like Loye, believes that such findings suggest that there are several potential futures, or holoverses, forming in the gathering mists of fate. But like other past-life researchers he also believes we create our own destiny, both individually and collectively, and thus the four scenarios are really a glimpse into the various potential futures the human race is creating for itself en masse.

Consequently, Snow recommends that instead of building bomb shelters or moving to areas that won't be destroyed by the "coming Earth changes" predicted by some psychics, we should spend time believing in and visualizing a positive future. He cites the Planetary Commission—the ad hoc collection of millions of individuals around the world who have agreed to spend the hour of 12:00 to 1:00 P.M., Greenwich mean time, each December thirty-first united in prayer and meditation on world peace and healing—as a step in the right direction. "If we are continually shaping our future physical reality by today's collective thoughts and actions, then the time to wake up to the alternative we have created is *now*," states Snow. "The choices between the kind of Earth represented by each of the Types are clear. Which do we want for our grandchildren? Which do we want perhaps to return to ourselves someday?"[89]

CHANGING THE PAST

The future may not be the only thing that can be formed and reshaped by human thought. At the 1988 Annual Convention of the Parapsychological Association, Helmut Schmidt and Marilyn Schlitz announced that several experiments they had conducted indicated the mind may be able to alter the past as well. In one study Schmidt and Schlitz used a computerized randomization process to record 1,000 different sequences of sound. Each sequence consisted of 100 tones of varying duration, some of them pleasing to the ear and some just bursts of noise. Because the selection process was random, according to the laws of probability each sequence should contain roughly 50 percent pleasing sounds and 50 percent noise.

Cassette recordings of the sequences were then mailed to volunteers. While listening to the prerecorded cassettes the subjects were told to try to psychokinetically increase the duration of the pleasing sounds and decrease the durations of the noise. After the subjects completed the task, they notified the lab of their attempts, and Schmidt and Schlitz then examined the original sequences. They discovered that the recordings the subjects listened to contained significantly longer stretches of pleasing sounds than noise. In other words, it appeared that the subjects had psychokinetically reached back through time and had an effect on the randomized process from which their *prerecorded* cassettes had been made.

In another test Schmidt and Schlitz programmed the computer to produce 100-tone sequences randomly composed of four different notes, and subjects were instructed to try to psychokinetically cause more high notes to appear on the tapes than low. Again a retroactive PK effect was found. Schmidt and Schlitz also discovered that volunteers who meditated regularly exerted a greater PK effect than nonmeditators, suggesting again that contact with the unconscious is the key to accessing the reality-structuring portions of the psyche.[90]

The idea that we can psychokinetically alter events that have already occurred is an unsettling notion, for we are so deeply programmed to believe the past is frozen as if it were a butterfly in glass, it is difficult for us to imagine otherwise. But in a holographic universe, a universe in which time is an illusion and reality is no more than a mind-created image, it is a possibility to which we may have to become accustomed.

A WALK THROUGH THE GARDEN OF TIME

As fantastic as the above two notions are, they are small change compared to the last category of time anomaly that merits our attention. On August 10, 1901, two Oxford professors, Anne Moberly, the principal of St. Hugh's College, Oxford, and Eleanor Jourdain, the vice principal, were walking through the garden of the Petit Trianon at Versailles when they saw a shimmering effect pass over the landscape in front of them, not unlike the special effects in a movie when it changes from one scene to another. After the shimmering passed they noticed that the landscape had changed. Suddenly the people around them were wearing eighteenth-century costumes and wigs and were behaving in an agitated manner. As the two women stood dumbfounded, a repulsive man with a pockmarked face approached and urged them to change their direction. They followed him past a line of trees to a garden where they heard strains of music floating through the air and saw an aristocratic lady painting a watercolor.

Eventually the vision vanished and the landscape returned to normal, but the transformation had been so dramatic that when the women looked behind them they realized the path they had just walked down was now blocked by an old stone wall. When they returned to England, they searched through historical records and concluded that they had been transported back in time to the day in which the sacking of the Tuileries and the massacre of the Swiss Guards had taken place—which accounted for the agitated manner of the people in the garden—and that the woman in the garden was none other than Marie Antoinette. So vivid was the experience that the women filled a book-length manuscript about the occurrence and presented it to the British Society for Psychical Research.[91]

What makes Moberly and Jourdain's experience so significant is that they did not simply have a retrocognitive vision of the past, but actually *walked back into the past*, meeting people and wandering around in the Tuileries garden as it was more than one hundred years earlier. Moberly and Jourdain's experience is difficult to accept as real, but given that it provided them with no obvious benefit, and most certainly put their academic reputations at risk, one is hard pressed to imagine what would motivate them to make up such a story.

And it is not the only such occurrence at the Tuileries to be reported to the British Society for Psychical Research. In May 1955, a London solicitor and his wife also encountered several eighteenth-century figures in the garden. And on another occasion, the staff of an embassy whose offices overlook Versailles claims to have watched the garden revert back to an earlier period of history as well.[92] Here in the United States parapsychologist Gardner Murphy, a former president of both the American Psychological Association and the American Society for Psychical Research, investigated a similar case in which a woman identified only by the name Buterbaugh looked out the window of her office at Nebraska Wesleyan University and saw the campus as it was fifty years earlier. Gone were the bustling streets and the sorority houses, and in their place was an open field and a sprinkling of trees, their leaves aflutter in the breeze of a summer long since passed.[93]

Is the boundary between the present and the past so flimsy that we can, under the right circumstances, stroll back into the past with the same ease that we can stroll through a garden? At present we simply do not know, but in a world that is comprised less of solid objects traveling in space and time, and more of ghostly holograms of energy sustained by processes that are at least partially connected to human consciousness, such events may not be as impossible as they appear.

And if this seems disturbing—this idea that our minds and even our bodies are far less bound by the strictures of time than we have previously imagined—we should remember that the idea the Earth is round once proved equally frightening to a humanity convinced that it was flat. The evidence presented in this chapter suggests that we are still children when it comes to understanding the true nature of time. And like all children poised on the threshold of adulthood, we should put aside our fears and come to terms with the way the world really is. For in a holographic universe, a universe in which all things are just ghostly coruscations of energy, more than just our understanding of time must change. There are still other shimmerings to cross our landscape, still deeper depths to plumb.

8

Traveling in the Superhologram

Access to holographic reality becomes *experientially* available when one's consciousness is freed from its dependence on the physical body. So long as one remains tied to the body and its sensory modalities, holographic reality *at best* can only be an intellectual construct. When one [is freed from the body] one experiences it directly. That is why mystics speak about their visions with such certitude and conviction, while those who haven't experienced this realm for themselves are left feeling skeptical or even indifferent.

—Kenneth Ring, Ph.D.
Life at Death

Time is not the only thing that is illusory in a holographic universe. Space, too, must be viewed as a product of our mode of perception. This is even more difficult to comprehend than the idea that time is a construct, for when it comes to trying to conceptualize "spacelessness" there are no easy analogies, no images of amoeboid universes or crystallizing futures, to fall back on. We are so conditioned to think in terms of space as an absolute that it is hard for us even to begin to imagine what it would be like to exist in a realm in which space did not exist. Nonetheless, there is evidence that we are ultimately no more bound by space than we are by time.

One powerful indication that this is so can be found in out-of-body phenomena, experiences in which an individual's conscious awareness appears to detach itself from the physical body and travel to some other location. Out-of-body experiences, or OBEs, have been reported throughout history by individuals from all walks of life. Aldous Huxley, Goethe, D. H. Lawrence, August Strindberg, and Jack London all reported having OBEs. They were known to the Egyptians, the North American Indians, the Chinese, the Greek philosophers, the medieval alchemists, the Oceanic peoples, the Hindus, the Hebrews, and the Moslems. In a cross-cultural study of 44 non-Western societies, Dean Shiels found that only three did *not* hold a belief in OBEs.[1] In a similar study anthropologist Erika Bourguignon looked at 488 world societies—or roughly 57 percent of all known societies—and found that 437 of them, or 89 percent, had at least some tradition regarding OBEs.[2]

Even today studies indicate that OBEs are still widespread. The late Dr. Robert Crookall, a geologist at the University of Aberdeen and an amateur parapsychologist, investigated enough cases to fill nine books on the subject. In the 1960s Celia Green, the director of the Institute of Psychophysical Research in Oxford, polled 115 students at Southampton University and found that 19 percent admitted to having an OBE. When 380 Oxford students were similarly questioned, 34 percent answered in the affirmative.[3] In a survey of 902 adults Haraldsson found that 8 percent had experienced being out of their bodies at least once in their life.[4] And a 1980 survey conducted by Dr. Harvey Irwin at the University of New England in Australia revealed that 20 percent of 177 students had experienced an OBE.[5] When averaged, these figures indicate that roughly one out of every five people will have an OBE at some point in his or her life. Other studies suggest the incidence may be closer to one in ten, but the fact remains: OBEs are far more common than most people realize.

The typical OBE is usually spontaneous and occurs most often during sleep, meditation, anesthesia, illness, and instances of traumatic pain (although they can occur under other circumstances as well). Suddenly a person experiences the vivid sensation that his mind has separated from his body. Frequently he finds himself floating over his body and discovers he can travel or fly to other locations. What is it like to find oneself free from the physical and staring down at one's own body? In a 1980 study of 339 cases of out-of-body travel, Dr. Glen

Gabbard of the Menninger Foundation in Topeka, Dr. Stuart Twemlow of the Topeka Veterans' Administration Medical Center, and Dr. Fowler Jones of the University of Kansas Medical Center found that a whopping 85 percent described the experience as pleasant and over half of them said it was joyful.[6]

I know the feeling. I had a spontaneous OBE as a teenager, and after recovering from the shock of finding myself floating over my body and staring down at myself asleep in bed, I had an indescribably exhilarating time flying through walls and soaring over the treetops. During the course of my bodiless journey I even stumbled across a library book a neighbor had lost and was able to tell her where the book was located the next day. I describe this experience in detail in *Beyond the Quantum.*

It is of no small significance that Gabbard, Twemlow, and Jones also studied the psychological profile of OBEers and found that they were psychologically normal and were on the whole extremely well adjusted. At the 1980 meeting of the American Psychiatric Association they presented their conclusions and told their colleagues that reassurances that OBEs are common occurrences and referring the patient to books on the subject may be "more therapeutic" than psychiatric treatment. They even hinted that patients might gain more relief by talking to a yogi than to a psychiatrist![7]

Such facts notwithstanding, no amount of statistical findings are as convincing as actual accounts of such experiences. For example, Kimberly Clark, a hospital social worker in Seattle, Washington, did not take OBEs seriously until she encountered a coronary patient named Maria. Several days after being admitted to the hospital Maria had a cardiac arrest and was quickly revived. Clark visited her later that afternoon expecting to find her anxious over the fact that her heart had stopped. As she had expected, Maria was agitated, but not for the reason she had anticipated.

Maria told Clark that she had experienced something very strange. After her heart had stopped she suddenly found herself looking down from the ceiling and watching the doctors and the nurses working on her. Then something over the emergency room driveway distracted her and as soon as she "thought herself" there, she *was* there. Next Maria "thought her way" up to the third floor of the building and found herself "eyeball to shoelace" with a tennis shoe. It was an old shoe and she noticed that the little toe had worn a hole through the fabric. She also noticed several other details, such as the fact that the lace was stuck under the heel. After Maria finished her account she begged Clark to please go to the ledge and see if there was a shoe there so that she could confirm whether her experience was real or not.

Skeptical but intrigued, Clark went outside and looked up at the ledge, but saw nothing. She went up to the third floor and began going in and out of patients' rooms looking through windows so narrow she had to press her face against the glass just to see the ledge at all. Finally, she found a room where she pressed her face against the glass and looked down and saw the tennis shoe. Still, from her vantage point she could not tell if the little toe had worn a place in the shoe or if any of the other details Maria had described were correct. It wasn't until she retrieved the shoe that she confirmed Maria's various observations. "The only way she would have had such a perspective was if she had been floating right outside and at very close range to the tennis shoe," states Clark, who has since become a believer in OBEs. "It was very concrete evidence for me."[8]

Experiencing an OBE during cardiac arrest is relatively common, so common that Michael B. Sabom, a cardiologist and professor of medicine at Emory University and a staff physician at the Atlanta Veterans' Administration Medical Center, got tired of hearing his patients recount such "fantasies" and decided to settle the matter once and for all. Sabom selected two groups of patients, one composed of 32 seasoned cardiac patients who had reported OBEs during their heart attacks, and one made up of 25 seasoned cardiac patients who had never experienced an OBE. He then interviewed the patients, asking the OBEers to describe their own resuscitation as they had witnessed it from the out-of-body state, and asking the nonexperiencers to describe what they imagined must have transpired during their resuscitation.

Of the nonexperiencers, 20 made major mistakes when they described their resuscitations, 3 gave correct but general descriptions, and 2 had no idea at all what had taken place. Among the experiencers, 26 gave correct but general descriptions, 6 gave highly detailed and accurate descriptions of their own resuscitation, and 1 gave a blow-by-blow accounting so accurate that Sabom was stunned. The results inspired him to delve even deeper into the phenomenon, and like Clark, he has now become an ardent believer and lectures widely on the subject. There appears "to be no plausible explanation for the accuracy of these observations involving the usual physical senses," he says. "The out-of-body hypothesis simply seems to fit best with the data at hand."[9]

Although the OBEs experienced by such patients are spontaneous, some people have mastered the ability well enough to leave their body at will. One of the most famous of these individuals is a former radio and television executive named Robert Monroe. When Monroe had his first OBE in the late 1950s he thought he was going crazy and immediately sought medical treatment. The doctors he consulted found nothing wrong, but he continued to have his strange experiences and continued to be greatly disturbed by them. Finally, after learning from a psychologist friend that Indian yogis reported leaving their bodies all the time, he began to accept his uninvited talent. "I had two options," Monroe recalls. "One was sedation for the rest of my life; the other was to learn something about this state so I could control it."[10]

From that day forward Monroe began keeping a written journal of his experiences, carefully documenting everything he learned about the out-of-body state. He discovered he could pass through solid objects and travel great distances in the twinkling of an eye simply by "thinking" himself there. He found that other people were seldom aware of his presence, although the friends whom he traveled to see while in this "second state" quickly became believers when he accurately described their dress and activity at the time of his out-of-body visit. He also discovered that he was not alone in his pursuit and occasionally bumped into other disembodied travelers. Thus far he has catalogued his experiences in two fascinating books, *Journeys Out of the Body* and *Far Journeys*.

OBEs have also been documented in the lab. In one experiment, parapsychologist Charles Tart was able to get a skilled OBEer he identifies only as Miss Z to identify correctly a five-digit number written on a piece of paper that could only be reached if she were floating in the out-of-body state.[11] In a series of experiments conducted at the American Society for Psychical Research in New York, Karlis Osis and psychologist Janet Lee Mitchell found several gifted subjects who were able to "fly in" from various locations around the country and correctly describe a wide range of target images, including objects placed on a table, colored geometric patterns placed on a free-floating shelf near the ceiling, and optical illusions that could only be seen when an observer peered through a small window in a special device.[12] Dr. Robert Morris, the director of research at the Psychical Research

Foundation in Durham, North Carolina, has even used animals to detect out-of-body visitations. In one experiment, for instance, Morris found that a kitten belonging to a talented out-of-body subject named Keith Harary consistently stopped meowing and started purring whenever Harary was invisibly present.[13]

OBEs as a Holographic Phenomenon

Considered as a whole the evidence seems unequivocal. Although we are taught that we "think" with our brains, this is not always true. Under the right circumstances our consciousness—the thinking, perceiving part of us—can detach from the physical body and exist just about anywhere it wants to. Our current scientific understanding cannot account for this phenomenon, but it becomes much more tractable in terms of the holographic idea.

Remember that in a holographic universe, location is itself an illusion. Just as an image of an apple has no specific location on a piece of holographic film, in a universe that is organized holographically things and objects also possess no definite location; everything is ultimately nonlocal, including consciousness. Thus, although our consciousness appears to be localized in our heads, under certain conditions it can just as easily appear to be localized in the upper corner of the room, hovering over a grassy lawn, or floating eyeball-to-shoelace with a tennis shoe on the third-floor ledge of a building.

If the idea of a nonlocal consciousness seems difficult to grasp, a useful analogy can once again be found in dreaming. Imagine that you are dreaming you are attending a crowded art exhibit. As you wander among the people and gaze at the artworks, your consciousness appears to be localized in the head of the person you are in the dream. But where is your consciousness really? A quick analysis will reveal that it is actually in everything in the dream, in the other people attending the exhibit, in the artworks, even in the very space of the dream. In a dream, location is also an illusion because everything—people, objects, space, consciousness, and so on—is unfolding out of the deeper and more fundamental reality of the dreamer.

Another strikingly holographic feature of the OBE is the plasticity of the form a person assumes once they are out of the body. After detaching from the physical, OBEers sometimes find themselves in a ghostlike body that is an exact replica of their biological body. This caused some researchers in the past to postulate that human beings possess a "phantom double" not unlike the doppelgänger of literature.

However, recent findings have exposed problems with this assumption. Although some OBEers describe this phantom double as naked, others find themselves in bodies that are fully clothed. This suggests that the phantom double is not a permanent energy replica of the biological body, but is instead a kind of hologram that can assume many shapes. This notion is borne out by the fact that phantom doubles are not the only forms people find themselves in during OBEs. There are numerous reports where people have also perceived themselves as balls of light, shapeless clouds of energy, and even no discernible form at all.

There is even evidence that the form a person assumes during an OBE is a direct consequence of their beliefs and expectations. For example, in his 1961 book *The Mystical Life*, mathematician J. H. M. Whiteman revealed that he experienced at least two OBEs a month during most of his adult life and recorded over two thousand such incidents. He also disclosed that he always felt like a woman trapped in a man's body, and during separation this sometimes resulted in his finding himself in female form. Whiteman experienced various other forms as well during his OB adventures, including children's bodies, and concluded that beliefs, both conscious and unconscious, were the determining factors in the form this second body assumed.[14]

Monroe agrees and asserts that it is our "thought habits" that create our OB forms. Because we are so habituated to being in a body, we have a tendency to reproduce the same form in the OB state. Similarly, he believes it is the discomfort most people feel when they are naked that causes OBEers to unconsciously sculpt clothing for themselves when they assume a human form. "I suspect that one may modify the Second Body into whatever form is desired," says Monroe.[15]

What is our true form, if any, when we are in the disembodied state? Monroe has found that once we drop all such disguises, we are at heart a "vibrational pattern [comprised] of many interacting and resonating frequencies."[16] This finding is also remarkably suggestive that something holographic is going on and offers further evidence that we—like all things in a holographic universe—are ultimately a frequency phenomenon which our mind converts into various holographic forms. It also adds credence to Hunt's conclusion that our consciousness is contained, not in the brain, but in a plasmic holographic energy field that both permeates and surrounds the physical body.

The form we assume while in the OB state is not the only thing that displays this holographic plasticity. Despite the accuracy of the observations made by talented OB travelers during their disembodied jaunts, researchers have long been troubled by some of the glaring inaccuracies that crop up as well. For instance, the title of the lost library book I stumbled across during my own OBE looked bright green while I was in a disembodied state. But after I was back in my physical body and returned to retrieve the book I saw that the lettering was actually black. The literature is filled with accounts of similar discrepancies, instances in which OB travelers accurately described a distant room full of people, save that they added an extra person or perceived a couch where there was really a table.

In terms of the holographic idea, one explanation may be that such OB travelers have not yet fully developed the ability to convert the frequencies they perceive while in a disembodied state into a completely accurate holographic representation of consensus reality. In other words, since OBEers appear to be relying on a completely new set of senses, these senses may still be wobbly and not yet proficient at the art of converting the frequency domain into a seemingly objective construct of reality.

These nonphysical senses are further hampered by the constraints our own self-limiting beliefs place upon them. A number of talented OB travelers have noted that once they became more at home in their second body they discovered that they could "see" in all directions at once without turning their heads. In other words, although seeing in all directions appears to be normal during the OB state, they were so accustomed to believing that they could see only through their eyes—even when they were in a nonphysical hologram of their body—that this belief at first kept them from realizing that they possessed 360-degree vision.

There is evidence that even our physical senses have fallen victim to this censorship. Despite our unwavering conviction that we see with our eyes, reports persist of individuals who possess "eyeless sight," or the ability to see with other areas of their bodies. Recently David Eisenberg, M.D., a clinical research fellow at the Harvard Medical School, published an account of two school-age Chinese sisters in Beijing who can "see" well enough with the skin in their armpits to read notes and identify colors.[17] In Italy the neurologist Cesare Lombroso studied a blind girl who could see with the tip of her nose and the lobe of her left ear.[18] In the 1960s the prestigious Soviet Academy of Science investigated a Russian peasant woman named Rosa Kuleshova, who could see photographs and read newspapers with the tips of her fingers, and pronounced her abilities genuine. Significantly, the Soviets ruled out the possibility that Kuleshova was simply detecting the varying amounts of stored heat different colors emanate naturally—Kuleshova could read a black and white newspaper *even when it was covered with a sheet of heated glass.* [19] Kuleshova became so renowned for her abilities that *Life* magazine eventually published an article about her.[20]

In short, there is evidence that we too are not limited to seeing only through our physical eyes. This is, of course, the message inherent in my father's friend Tom's ability to read the inscription on a watch even when it was shielded by his daughter's stomach, and also in the remote-viewing phenomenon. One cannot help but wonder if eyeless sight is actually just further evidence that reality is indeed *maya,* an illusion, and our physical body, as well as all the seeming absoluteness of its physiology, is as much a holographic construct of our perception as our second body. Perhaps we are so deeply habituated to believing that we can see only through our eyes that even in the physical we have shut ourselves off from the full range of our perceptual capabilities.

Another holographic aspect of OBEs is the blurring of the division between past and future that sometimes occurs during such experiences. For example, Osis and Mitchell discovered that when Dr. Alex Tanous, a well-known psychic and talented OB traveler from Maine, flew in and attempted to describe the test objects they placed on a table, he had a tendency to describe items that were placed there days *later!*[21] This suggests that the realm people enter during the OB state is one of the subtler levels of reality Bohm speaks about, a region that is closer to the implicate and hence closer to the level of reality in which the division between past, present, and future ceases to exist. Put another way, it appears that instead of tuning into the frequencies that encode the present, Tanous's mind inadvertently tuned into frequencies that contained information about the future and converted those into a hologram of reality.

That Tanous's perception of the room was a holographic phenomenon and not just a precognitive vision that took place solely in his head is underscored by another fact. The day of his schedule to produce an

OBE Osis asked New York psychic Christine Whiting to hold vigil in the room and try to describe any projector she might "see" visiting there. Despite Whiting's ignorance of who would be flying in or when, when Tanous made his OB visit she saw his apparition clearly and described him as wearing brown corduroy pants and a white cotton shirt, the clothing Dr. Tanous was wearing in Maine at the time of his attempt.[22]

Harary has also made occasional OB journeys into the future and agrees that the experiences are qualitatively different from other precognitive experiences. "OBEs to future time and space differ from regular precognitive dreams in that I am definitely 'out' and moving through a black, dark area that ends at some lighted future scene," he states. When he makes an OB visit to the future he has sometimes even seen a silhouette of his future self in the scene, and this is not all. When the events he has witnessed eventually come to pass, *he can also sense his time-traveling OB self in the actual scene with him.* He describes this eerie sensation as "meeting myself 'behind' myself as if I were two beings," an experience that surely must put normal déjà vus to shame.[23]

There are also cases on record of OB journeys into the past. The Swedish playwright August Strindberg, himself a frequent OB traveler, describes one in his book *Legends.* The occurrence took place while Strindberg was sitting in a wine shop, trying to persuade a young friend not to give up his military career. To bolster his argument Strindberg brought up a past incident involving both of them that had taken place one evening in a tavern. As the playwright proceeded to describe the event he suddenly "lost consciousness" only to find himself sitting in the tavern in question and reliving the occurrence. The experience lasted only for a few moments, and then he abruptly found himself back in his body and in the present.[24] The argument can also be made that the retrocognitive visions we examined in the last chapter in which clairvoyants had the experience that they were actually present during, and even "floating" over, the historical scenes they were describing are also a form of OB projection into the past.

Indeed, when one reads the voluminous literature now available on the OB phenomenon, one is repeatedly struck at the similarities between OB travelers' descriptions of their experiences and characteristics we have now come to associate with a holographic universe. In addition to describing the OB state as a place where time and space no longer properly exist, where thought can be transformed into hologramlike forms, and where consciousness is ultimately a pattern of vibrations, or frequencies, Monroe notes that perception during OBEs seems based less on "a reflection of light waves" and more on "an impression of radiation," an observation that suggests once again that when one enters the OB realm one begins to enter Pribram's frequency domain.[25] Other OB travelers have also referred to the frequencylike quality of the Second State. For instance, Marcel Louis Forhan, a French OB experiencer who wrote under the name of "Yram," spends much of his book, *Practical Astral Projection*, trying to describe the wavelike and seemingly electromagnetic qualities of the OB realm. Still others have commented on the sense of cosmic unity one experiences during the state and have summarized it as a feeling that "everything is everything," and "I am that."[26]

As holographic as the OBE is, it is only the tip of the iceberg when it comes to more direct experience of the frequency aspects of reality. Although OBEs are only experienced by a segment of the human race, there is another circumstance under which we all come into closer contact with the frequency domain. That is when we journey to that undiscovered country from whose bourn no traveler returns. The rub, with all due respect to Shakespeare, is that some travelers *do* return. And the stories they tell are filled with features that smack once again of things holographic.

The Near-Death Experience

By now, nearly everyone has heard of near-death experiences, or NDEs, incidents in which individuals are declared clinically "dead," are resuscitated, and report that during the experience they left their physical body and visited what appeared to be the realm of the afterlife. In our own culture NDEs first came to prominence in 1975 when Raymond A. Moody, Jr., a psychiatrist who also has a Ph.D. in philosophy, published his best-selling investigation of the subject, *Life after Life*. Shortly thereafter Elisabeth Kubler-Ross revealed that she had simultaneously conducted similar research and had duplicated Moody's findings. Indeed, as more and more researchers began to document the phenomenon it became increasingly clear that NDEs were not only incredibly widespread—a 1981 Gallup poll found that eight million adult Americans had experienced an NDE, or roughly one person in twenty—but provided the most compelling evidence to date for survival after death.

Like OBEs, NDEs appear to be a universal phenomenon. They are described at length in both the eighth-century Tibetan Book of the Dead and the 2,500-year-old Egyptian Book of the Dead. In Book X of *The Republic* Plato gives a detailed account of a Greek soldier named Er, who came alive just seconds before his funeral pyre was to be lit and said that he had left his body and went through a "passageway" to the land of the dead. The Venerable Bede gives a similar account in his eighth-century work *A History of the English Church and People*, and, in fact, in her recent book *Otherworld Journeys* Carol Zaleski, a lecturer on the study of religion at Harvard, points out that medieval literature is filled with accounts of NDEs.

NDEers also have no unique demographic characteristics. Various studies have shown that there is no relationship between NDEs and a person's age, sex, marital status, race, religion and/or spiritual beliefs, social class, educational level, income, frequency of church attendance, size of home community, or area of residence. NDEs, like lightning, can strike anyone at any time. The devoutly religious are no more likely to have an NDE than nonbelievers.

One of the most interesting aspects of the ND phenomenon is the consistency one finds from experience to experience. A summary of a typical NDE is as follows:

A man is dying and suddenly finds himself floating above his body and watching what is going on. Within moments he travels at great speed through a darkness or a tunnel. He enters a realm of dazzling light and is warmly met by recently deceased friends and relatives. Frequently he hears indescribably beautiful music and sees sights—rolling meadows, flower-filled valleys, and sparkling streams—more lovely than anything he has seen on earth. In this light-filled world he feels no pain or fear and is pervaded with an overwhelming feeling of joy, love, and peace. He meets a "being (and or beings) of light" who emanates a feeling of enormous compassion, and is prompted by the being(s) to experience a "life review," a panoramic replay of his life. He becomes so enraptured by his experience of this greater reality that he desires nothing more than to stay. However, the being tells him that it is not his time yet and persuades him to return to his earthly life and reenter his physical body.

It should be noted this is only a general description and not all NDEs contain all of the elements described. Some may lack some of the above-mentioned features, and others may contain additional ingredients. The symbolic trappings of the experiences can also vary. For example, although NDEers in Western cultures tend to enter the realm of the afterlife by passing through a tunnel, experiencers from other cultures might walk down a road or pass over a body of water to arrive in the world beyond.

Nevertheless, there is an astonishing degree of agreement among the NDEs reported by various cultures throughout history. For instance, the life review, a feature that crops up again and again in modern-day NDEs, is also described in the Tibetan Book of the Dead, the Egyptian Book of the Dead, in Plato's account of what Er experienced during his sojourn in the hereafter, and in the 2,000-year-old yogic writings of the Indian sage Patanjali. The cross-cultural similarities between NDEs has also been confirmed in formal study. In 1977, Osis and Haraldsson compared nearly nine hundred deathbed visions reported by patients to doctors and other medical personnel in both India and the United States and found that although there were various cultural differences—for example, Americans tended to view the being of light as a Christian religious personage and Indians perceived it to be a Hindu one—the "core" of the experience was substantially the same and resembled the NDEs described by Moody and Kubler-Ross.[27]

Although the orthodox view of NDEs is that they are just hallucinations, there is substantial evidence that this is not the case. As with OBEs, when NDEers are out-of-body, they are able to report details they have no normal sensory means of knowing. For example, Moody reports a case in which a woman left her body during surgery, floated into the waiting room, and saw that her daughter was wearing mismatched plaids. As it turned out, the maid had dressed the little girl so hastily she had not noticed the error and was astounded when the mother, who did not physically see the little girl that day, commented on the fact.[28] In another case, after leaving her body, a female NDEer went to the hospital lobby and overheard her brother-in-law tell a friend that it looked like he was going to have to cancel a business trip and instead be one of his sister-in-law's pallbearers. After the woman recovered, she reprimanded her astonished brother-in-law for writing her off so quickly.[29]

And these are not even the most extraordinary examples of sensory awareness in the ND out-of-body state. NDE researchers have found that even patients who are blind, and have had no light perception for years, can see and accurately describe what is going on around them when they have left their bodies during an NDE. Kubler-Ross has encountered several such individuals and has interviewed them at length to determine their accuracy. "To our amazement, they were able to describe the color and design of clothing and jewelry the people present wore," she states.[30]

Most staggering of all are those NDEs and deathbed visions involving two or more individuals. In one case, as a female NDEer found herself moving through the tunnel and approaching the realm of light, she saw a friend of hers coming back! As they passed, the friend telepathically communicated to her that he had died, but was being "sent back." The woman, too, was eventually "sent back" and after she recovered she discovered that her friend had suffered a cardiac arrest at approximately the same time of her own experience.[31]

There are numerous other cases on record in which dying individuals knew who was waiting for them in the world beyond before news of the person's death arrived through normal channels.[32]

And if there is still any doubt, yet another argument against the idea that NDEs are hallucinations is their occurrence in patients who have flat EEGs. Under normal circumstances whenever a person talks, thinks, imagines, dreams, or does just about anything else, their EEG registers an enormous amount of activity. Even hallucinations measure on the EEG. But there are many cases in which people with flat EEGs have had NDEs. Had their NDEs been simple hallucinations, they would have registered on their EEGs.

In brief, when all these facts are considered together—the widespread nature of the NDE, the absence of demographic characteristics, the universality of the core experience, the ability of NDEers to see and know things they have no normal sensory means of seeing and knowing, and the occurrence of NDEs in patients who have flat EEGs—the conclusion seems inescapable: People who have NDEs are not suffering from hallucinations or delusional fantasies, *but are actually making visits to an entirely different level of reality.*

This is also the conclusion reached by many NDE researchers. One such researcher is Dr. Melvin Morse, a pediatrician in Seattle, Washington. Morse first became interested in NDEs after treating a seven-year-old drowning victim. By the time the little girl was resuscitated she was profoundly comatose, had fixed and dilated pupils, no muscle reflexes, and no corneal response. In medical terms this gave her a Glascow Coma Score of three, indicating that she was in a coma so deep she had almost no chance of ever recovering. Despite these odds, she made a full recovery and when Morse looked in on her for the first time after she regained consciousness she recognized him and said that she had watched him working on her comatose body. When Morse questioned her further she explained that she had left her body and passed through a tunnel into heaven where she had met "the Heavenly Father." The Heavenly Father told her she was not really meant to be there yet and asked if she wanted to stay or go back. At first she said she wanted to stay, but when the Heavenly Father pointed out that that decision meant she would not be seeing her mother again, she changed her mind and returned to her body.

Morse was skeptical but fascinated and from that point on set out to learn everything he could about NDEs. At the time, he worked for an air transport service in Idaho that carried patients to the hospital, and this afforded him the opportunity to talk with scores of resuscitated children. Over a ten-year period he interviewed every child survivor of cardiac arrest at the hospital, and over and over they told him the same thing. After going unconscious they found themselves outside their bodies, watched the doctors working on them, passed through a tunnel, and were comforted by luminous beings.

Morse continued to be skeptical, and in his increasingly desperate search for some logical explanation he read everything he could find on the side effects of the drugs his patients were taking, and explored various psychological explanations, but nothing seemed to fit. "Then one day I read a long article in a medical journal that tried to explain NDEs as being various tricks of the brain," says Morse. "By then I had studied NDEs extensively and none of the explanations that this researcher listed made sense. It was finally clear to me that he had missed the most obvious explanation of all—NDEs are real. He had missed the possibility that the soul really does travel."[33]

Moody echoes the sentiment and says that twenty years of research have convinced him that NDEers have indeed ventured into another level of reality. He believes that most other NDE researchers feel the same. "I have talked to almost every NDE researcher in the world about his or her work. I know that most of them believe in their hearts that NDEs are a glimpse of life after life. But as scientists and people of medicine, they still haven't come up with 'scientific proof' that a part of us goes on living after our physical being is dead. This lack of proof keeps them from going public with their true feelings."[34]

As a result of his 1981 survey, even George Gallup, Jr., the president of the Gallup Poll, agrees: "A growing number of researchers have been gathering and evaluating the accounts of those who have had strange near-death encounters. And the preliminary results have been highly suggestive of some sort of encounter with an extradimensional realm of reality. Our own extensive survey is the latest in these studies and is also uncovering some trends that point toward a super parallel universe of some sort."[35]

A Holographic Explanation of the Near-Death Experience

These are astounding assertions. What is even more astounding is that the scientific establishment has for the most part ignored both the conclusions of these researchers and the vast body of evidence that compels them to make such statements. The reasons for this are complex and varied. One is that it is currently not fashionable in science to consider seriously any phenomenon that seems to support the idea of a spiritual reality, and, as mentioned at the beginning of this book, beliefs are like addictions and do not surrender their grip easily. Another reason, as Moody mentions, is the widespread prejudice among scientists that the only ideas that have any value or significance are those that can be proven in a strict scientific sense. Yet another is the inability of our current scientific understanding of reality even to begin to explain NDEs if they are real.

This last reason, however, may not be the problem it seems. Several NDE researchers have pointed out that the holographic model offers us a way to understand these experiences. One such researcher is Dr. Kenneth Ring, a professor of psychology at the University of Connecticut and one of the first NDE researchers to use statistical analysis and standardized interviewing techniques to study the phenomenon. In his 1980 book *Life at Death*, Ring spends considerable time arguing in favor of a holographic explanation of the NDE. Put bluntly, Ring believes that NDEs are also ventures into the more frequency-like aspects of reality.

Ring bases his conclusion on the numerous suggestively holographic aspects of the NDE. One is the tendency of experiencers to describe the world beyond as a realm composed of "light," "higher vibrations," or "frequencies." Some NDEers even refer to the celestial music that often accompanies such experiences as more "a combination of vibrations" than actual sounds—observations that Ring believes are evidence that the act of dying involves a shift of consciousness away from the ordinary world of appearances and into a more holographic reality of pure frequency. NDEers also frequently say that the realm is suffused with a light more brilliant than any they have ever seen on earth, but one that, despite its unfathomable intensity, does not hurt the eyes, characterizations that Ring feels are further evidence of the frequency aspects of the hereafter.

Another feature Ring finds undeniably holographic is NDEers' descriptions of time and space in the afterlife realm. One of the most commonly reported characteristics of the world beyond is that it is a dimension in which time and space cease to exist. "I found myself in a space, in a period of time, I would say, where all space and time was negated," says one NDEer clumsily.[36] "It *has* to be out of time and space. It *must* be, because . . . it can't be put *into* a time thing," says another.[37] Given that time and space are collapsed and location has no meaning in the frequency domain, this is precisely what we would expect to find if NDEs take place in a holographic state of consciousness, says Ring.

If the near-death realm is even more frequencylike than our own level of reality, why does it appear to have any structure at all? Given that both OBEs and NDEs offer ample evidence that the mind can exist independently of the brain, Ring believes it is not too farfetched to assume that it, too, functions holographically. Thus, when the mind is in the "higher" frequencies of the near-death dimension, it continues to do what it does best, translate those frequencies into a world of appearances. Or as Ring puts it, "I believe that this is a realm that is created by *interacting thought structures*. These structures or 'thought-forms' combine to form patterns, just as interference waves form patterns on a holographic plate. And just as the holographic image appears to be fully real when illuminated by a laser beam, so the images produced by interacting thought-forms appear to be real."[38]

Ring is not alone in his speculations. In the keynote address for the 1989 meeting of the International Association for Near-Death Studies (IANDS), Dr. Elizabeth W. Fenske, a clinical psychologist in private

practice in Philadelphia, announced that she, too, believes that NDEs
are journeys into a holographic realm of higher frequencies. She
agrees with Ring's hypothesis that the landscapes, flowers, physical
structures, and so forth, of the afterlife dimension are fashioned out
of interacting (or interfering) thought patterns. "I think we've come
to the point in NDE research where it's difficult to make a distinction
between thought and light. In the near-death experience thought
seems to be light," she observes.[39]

Heaven as Hologram

In addition to those mentioned by Ring and Fenske, the NDE has
numerous other features that are markedly holographic. Like
OBEers, after NDEers have detached from the physical they find
themselves in one of two forms, either as a disembodied cloud of
energy, or as a hologramlike body sculpted by thought. When the
latter is the case, the mind-created nature of the body is often surprisingly obvious to the NDEer. For example, one near-death survivor
says that when he first emerged from his body he looked "something
like a jellyfish" and fell lightly to the floor like a soap bubble. Then he
quickly expanded into a ghostly three-dimensional image of a naked
man. However, the presence of two women in the room embarrassed
him and to his surprise, this feeling caused him suddenly to become
clothed (the women, however, never offered any indication that they
were able to see any of this).[40]

That our innermost feelings and desires are responsible for creating
the form we assume in the afterlife dimension is evident in the experiences of other NDEers. People who are confined in wheelchairs in
their physical existence find themselves in healthy bodies that can run
and dance. Amputees invariably have their limbs back. The elderly
often inhabit youthful bodies, and even stranger, children frequently
see themselves as adults, a fact that may reflect every child's fantasy
to be a grown-up, or more profoundly, may be a symbolic indication
that in our souls some of us are much older than we realize.

These hologramlike bodies can be remarkably detailed. In the incident involving the man who became embarrassed at his own nakedness, for example, the clothing he materialized for himself was so
meticulously wrought that he could even make out the seams in the

material![41] Similarly, another man who studied his hands while in the ND state said they were "composed of light with tiny structures in them" and when he looked closely he could even see "the delicate whorls of his fingerprints and tubes of light up his arms."[42]

Some of Whitton's research is also relevant to this issue. Amazingly, when Whitton hypnotized patients and regressed them to the between-life state, they too reported all the classic features of the NDE, passage through a tunnel, encounters with deceased relatives and/or "guides," entrance into a splendorous light-filled realm in which time and space no longer existed, encounters with luminous beings, and a life review. In fact, according to Whitton's subjects the main purpose of the life review was to refresh their memories so they could more mindfully plan their next life, a process in which the beings of light gently and noncoercively assisted.

Like Ring, after studying the testimony of his subjects Whitton concluded that the shapes and structures one perceives in the afterlife dimension are thought-forms created by the mind. "René Descartes' famous dictum, 'I think, therefore I am,' is never more pertinent than in the between-life state," says Whitton. "There is no experience of existence without thought."[43]

This was especially true when it came to the form Whitton's patients assumed in the between-life state. Several said they didn't even have a body unless they were thinking. "One man described it by saying that if he stopped thinking he was merely a cloud in an endless cloud, undifferentiated," he observes. "But as soon as he started to think, he became himself"[44] (a state of affairs that is oddly reminiscent of the subjects in Tart's mutual hypnosis experiment who discovered they didn't have hands unless they *thought* them into existence). At first the bodies Whitton's subjects assumed resembled the persons they had been in their last life. But as their experience in the between-life state continued, they gradually became a kind of hologramlike composite of all of their past lives.[45] This composite identity even had a name separate from any of the names they had used in their physical incarnations, although none of his subjects was able to pronounce it using their physical vocal cords.[46]

What do NDEers look like when they have not constructed a hologramlike body for themselves? Many say that they were not aware of any form and were simply "themselves" or "their mind." Others have more specific impressions and describe themselves as "a cloud of colors," "a mist," "an energy pattern," or "an energy field," terms that again suggest that we are all ultimately just frequency phenomena, patterns of some unknown vibratory energy enfolded in the greater matrix of the frequency domain. Some NDEers assert that in addition to being composed of colored frequencies of light, we are also constituted out of sound. "I realized that each person and thing has its own musical tone range as well as its own color range," says an Arizona housewife who had an NDE during childbirth. "If you can imagine yourself effortlessly moving in and out among prismatic rays of light and hearing each person's musical notes join and harmonize with your own when you touch or pass them, you would have some idea of the unseen world." The woman, who encountered many individuals in the afterlife realm who manifested only as clouds of colors and sound, believes the mellifluous tones each soul emanates are what people are describing when they say they hear beautiful music in the ND dimension.[47]

Like Monroe, some NDEers report being able to see in all directions at once while in the disembodied state. After wondering what he looked like, one man said he suddenly found himself staring at his own back.[48] Robert Sullivan, an amateur NDE researcher from Pennsylvania who specializes in NDEs by soldiers during combat, interviewed a World War II veteran who temporarily retained this ability even after he returned to his physical body. "He experienced three-hundred-sixty-degree vision while running away from a German machine-gun nest," says Sullivan. "Not only could he see ahead as he ran, but he could see the gunners trying to draw a bead on him from behind."[49]

Instantaneous Knowledge

Another part of the NDE that possesses many holographic features is the life review. Ring refers to it as "a holographic phenomenon par excellence." Grof and Joan Halifax, a Harvard medical anthropologist and the coauthor (with Grof) of *The Human Encounter with Death*, have also commented on the life review's holographic aspects. According to several NDE researchers, including Moody, even many NDEers themselves use the term "holographic" when describing the experience.[50]

The reason for this characterization is obvious as soon as one begins to read accounts of the life review. Again and again NDEers use the same adjectives to describe it, referring to it as an incredibly vivid, wrap-around, three-dimensional replay of their entire life. "It's like climbing right inside a movie of your life," says one NDEer. "Every moment from every year of your life is played back in complete sensory detail. Total, total recall. And it all happens in an instant."[51] "The whole thing was really odd. I was there; I was actually seeing these flashbacks; I was actually walking through them, and it was so fast. Yet, it was slow enough that I could take it all in," says another.[52]

During this instantaneous and panoramic remembrance NDEers reexperience all the emotions, the joys and the sorrows, that accompanied all of the events in their life. More than that, they feel all of the emotions of the people with whom they have interacted as well. They feel the happiness of all the individuals to whom they've been kind. If they have committed a hurtful act, they become acutely aware of the pain their victim felt as a result of their thoughtlessness. And no event seems too trivial to be exempt. While reliving a moment in her childhood, one woman suddenly experienced all the loss and powerlessness her sister had felt after she (then a child) snatched a toy away from her sister.

Whitton has uncovered evidence that thoughtless acts are not the only things that cause individuals remorse during the life review. Under hypnosis his subjects reported that failed dreams and aspirations—things they had hoped to accomplish during their life but had not—also caused them pangs of sadness.

Thoughts, too, are replayed with exacting fidelity during the life review. Reveries, faces glimpsed once but remembered for years, things that made one laugh, the joy one felt when gazing at a particular painting, childish worries, and long forgotten daydreams—all flit through one's mind in a second. As one NDEer summarizes, "Not even your thoughts are lost. . . . Every thought was there."[53]

And so, the life review is holographic not only in its three-dimensionality, but in the amazing capacity for information storage the process displays. It is also holographic in a third way. Like the kabbalistic "aleph," a mythical point in space and time that contains all other points in space and time, it is a moment that contains all other moments. Even the ability to perceive the life review seems holographic in that it is a faculty capable of experiencing something that is paradoxically at once both incredibly rapid and yet slow enough to witness in detail. As an NDEer in 1821 put it, it is the ability to "simultaneously comprehend the whole and every part."[54]

In fact, the life review bares a marked resemblance to the afterlife judgment scenes described in the sacred texts of many of the world's great religions, from the Egyptian to the Judeo-Christian, but with one crucial difference. Like Whitton's subjects, NDEers universally report that they are *never judged by the beings of light,* but feel only love and acceptance in their presence. *The only judgment that ever takes place is self-judgment and arises solely out of the NDEer's own feelings of guilt and repentance.* Occasionally the beings do assert themselves, but instead of behaving in an authoritarian manner, they act as guides and counselors whose only purpose is to teach.

This total lack of cosmic judgment and/or any divine system of punishment and reward has been and continues to be one of the most controversial aspects of the NDE among religious groups, but it is one of the most oft reported features of the experience. What is the explanation? Moody believes it is as simple as it is polemic. We live in a universe that is far more benevolent than we realize.

That is not to say that anything goes during the life review. Like Whitton's hypnotic subjects, after arriving in the realm of light NDEers appear to enter a state of heightened or metaconsciousness awareness and become lucidly honest in their self-reflections.

It also does not mean that the beings of light prescribe no values. In NDE after NDE they stress two things. One is the importance of love. Over and over they repeat this message, that we must learn to replace anger with love, learn to love more, learn to forgive and love everyone unconditionally, and learn that we in turn *are* loved. This appears to be the only moral criterion the beings use. Even sexual activity ceases to possess the moral stigma we humans are so fond of attaching to it. One of Whitton's subjects reported that after living several withdrawn and depressed incarnations he was urged to plan a life as an amorous and sexually active female in order to add balance to the overall development of his soul.[55] It appears that in the minds of the beings of light, compassion is the barometer of grace, and time and time again when NDEers wonder if some act they committed was right or wrong, the beings counter their inquiries only with a question: Did you do it out of love? Was the motivation love?

That is why we have been placed here on the earth, say the beings, to learn that love is the key. They acknowledge that it is a difficult undertaking, but intimate that it is crucial to both our biological and spiritual existence in ways that we have perhaps not even begun to fathom. Even children return from the near-death realm with this message firmly impressed in their thoughts. States one little boy who after being hit by a car was guided into the world beyond by two people in "very white" robes: "What I learned there is that the most important thing is loving while you are alive."[56]

The second thing the beings emphasize is knowledge. Frequently NDEers comment that the beings seemed pleased whenever an incident involving knowledge or learning flickered by during their life review. Some are openly counseled to embark on a quest for knowledge after they return to their physical bodies, especially knowledge related to self-growth or that enhances one's ability to help other people. Others are prodded with statements such as "learning is a continuous process and goes on even after death" and "knowledge is one of the few things you will be able to take with you after you have died."

The preeminence of knowledge in the afterlife dimension is apparent in another way. Some NDEers discovered that in the presence of the light they suddenly had direct access to *all* knowledge. This access manifested in several ways. Sometimes it came in response to inquiries. One man said that all he had to do was ask a question, such as what would it be like to be an insect, and instantly the experience was his.[57] Another NDEer described it by saying, "You can think of a question . . . and *immediately* know the answer to it. As simple as that. And it can be any question whatsoever. It can be on a subject that you don't know anything about, that you are not in the proper position even to understand and the light will give you the instantaneous correct answer and make you understand it."[58]

Some NDEers report that they didn't even have to ask questions in order to access this infinite library of information. Following their life review they just suddenly knew everything, all the knowledge there was to know from the beginning of time to the end. Others came into contact with this knowledge after the being of light made some specific gesture, such as wave its hand. Still others said that instead of acquiring the knowledge, they *remembered* it, but forgot most of what they recalled as soon as they returned to their physical bodies (an amnesia that seems to be universal among NDEers who are privy to such visions).[59] Whatever the case, it appears that once we are in the world beyond, it is no longer necessary to enter an altered state of consciousness in order to have access to the transpersonal and infinitely interconnected informational realm experienced by Grof's patients.

In addition to being holographic in all the ways already mentioned, this vision of total knowledge has another holographic characteristic. NDEers often say that during the vision the information arrives in "chunks" that register instantaneously in one's thoughts. In other words, rather than being strung out in a linear fashion like words in a sentence or scenes in a movie, all the facts, details, images, and pieces of information burst into one's awareness in an instant. One NDEer referred to these bursts of information as "bundles of thought."[60] Monroe, who has also experienced such instantaneous explosions of information while in the OB state, calls them "thought balls."[61]

Indeed, anyone who possesses any appreciable psychic ability is familiar with this experience, for this is the form in which one receives psychic information as well. For instance, sometimes when I meet a stranger (and on occasion even when I just hear a person's name), a thought ball of information about that person will instantly flash into my awareness. This thought ball can include important facts about the person's psychological and emotional makeup, their health, and even scenes from their past. I find that I am especially prone to getting thought balls about people who are in some kind of crisis. For example, recently I met a woman and instantly knew she was contemplating suicide. I also knew some of the reasons why. As I always do in such situations, I started talking to her and cautiously maneuvered the conversation to things psychic. After finding out that she was receptive to the subject, I confronted her with what I knew and got her to talk about her problems. I got her to promise to seek some kind of professional counseling instead of the darker option she was considering.

Receiving information in this manner is similar to the way one becomes aware of information while dreaming. Virtually everyone has had a dream in which they find themselves in a situation and suddenly know all kinds of things about it without being told. For instance, you might dream you are at a party and as soon as you are there you know who it is being given for and why. Similarly, everyone has had a detailed idea or inspiration dawn upon them in a flash. Such experiences are lesser versions of the thought ball effect.

Interestingly, because these bursts of psychic information arrive in nonlinear chunks, it sometimes takes me several moments to translate them into words. Like the psychological gestalts experienced by individuals during transpersonal experiences, they are holographic in the sense that they are instantaneous "wholes" our time-oriented minds must struggle with for a moment in order to unravel and convert into a serial arrangement of parts.

What form does the knowledge contained in the thought balls experienced during NDEs take? According to NDEers all forms of communication are used, sounds, moving hologramlike images, even telepathy—a fact that Ring believes demonstrates once again that the hereafter is "a world of existence where thought is king."[62]

The thoughtful reader may immediately wonder why the quest for learning is so important during life if we have access to all knowledge after we die? When asked this question NDEers replied that they weren't certain, but felt strongly that it had something to do with the purpose of life and the ability of each individual to reach out and help others.

Life Plans and Parallel Time Tracks

Like Whitton, NDE researchers have also uncovered evidence that our lives are planned beforehand, at least to some extent, and we each play a role in the creation of this plan. This is apparent in several aspects of the experience. Frequently after arriving in the world of light, NDEers are told that "it is not their time yet." As Ring points out, this remark clearly implies the existence of some kind of "life plan."[63] It is also clear that NDEers play a role in the formulation of these destinies, for they are often given the *choice* whether to return or stay. There are even instances of NDEers being told that it *is* their time and still being allowed to return. Moody cites a case in which a man started to cry when he realized he was dead because he was afraid his wife wouldn't be able to raise their nephew without him. On hearing this the being told him that since he wasn't asking for himself he would be allowed to return.[64] In another case a woman argued that she hadn't danced enough yet. Her remark caused the being of light to give a hearty laugh and she, too, was given permission to return to physical life.[65]

That our future is at least partially sketched out is also evident in a phenomenon Ring calls the "personal flashforward." On occasion, during the vision of knowledge, NDEers are shown glimpses of their own future. In one particularly striking case a child NDEer was told various specifics about his future, including the fact that he would be married at age twenty-eight and would have two children. He was even shown his adult self and his future children sitting in a room of the house he would eventually be living in, and as he gazed at the room he noticed something very strange on the wall, something that his mind could not grasp. Decades later and after each of these predictions had come to pass, he found himself in the very scene he had witnessed as a child and realized that the strange object on the wall was a "forced-air heater," a kind of heater that had not yet been invented at the time of his NDE.[66]

In another equally astonishing personal flashforward a female NDEer was shown a photograph of Moody, told his full name, and told that when the time was right she would tell him about her experience. The year was 1971 and Moody had not yet published *Life after Life*, so his name and picture meant nothing to the woman. However, the time became "right" four years later when Moody and his family unwittingly moved to the very street on which the woman lived. That Halloween Moody's son was out trick-or-treating and knocked on the woman's door. After hearing the boy's name, the woman told him to tell his father she had to talk to him, and when Moody obliged she related her remarkable story.[67]

Some NDEs even support Loye's proposal that several holographic parallel universes, or time tracks, exist. On occasion NDEers are shown personal flashforwards and told that the future they have witnessed will come to pass *only if they continue on their current path.* In one unique instance an NDEer was shown a completely different history of the earth, a history that would have developed if "certain events" had not taken place around the time of the Greek philosopher and mathematician Pythagoras three thousand years ago. The vision revealed that if these events, the precise nature of which the woman does not disclose, had failed to take place, we would now be living in a world of peace and harmony marked "by the absence of religious wars and of a Christ figure."[68] Such experiences suggest that the laws of time and space operative in a holographic universe may be very strange indeed.

Even NDEers who do not experience direct evidence of the role they play in their own destiny often come back with a firm understanding of the holographic interconnectedness of all things. As a sixty-two-year-old businessman who had an NDE during a cardiac arrest puts it, "One thing I learned was that we are all part of one big, living universe. If we think we can hurt another person or another living thing without hurting ourselves we are sadly mistaken. I look at a forest or a flower or a bird now, and say, 'That is me, part of me.' We are connected with all things and if we send love along those connections, then we are happy."[69]

You Can Eat but You Don't Have To

The holographic and mind-created aspects of the near-death dimension are apparent in myriad other ways. In describing the hereafter one child said that food appeared whenever she wished for it, but there was no need to eat, an observation that underscores once again the illusory and hologramlike nature of afterlife reality.[70] Even the symbolic language of the psyche is given "objective" form. For example, one of Whitton's subjects said that when he was introduced to a woman who was going to figure prominently in his next life, instead of appearing as a human she appeared as a shape that was half-rose, half-cobra. After being directed to figure out the meaning of the symbolism, he realized that he and the woman had been in love with one another in two other lifetimes. However, she had also twice been responsible for his death. Thus, instead of manifesting as a human, the loving and sinister elements of her character caused her to appear in a hologramlike form that better symbolized these two dramatically polar qualities.[71]

Whitton's subject is not alone in his experience. Hazrat Inayat Khan said that when he entered a mystical state and traveled to "divine realities," the beings he encountered also occasionally appeared in half-human, half-animal forms. Like Whitton's subject, Khan discerned that these transfigurations were symbolic, and when a being appeared as part animal it was because the animal symbolized some quality the being possessed. For example, a being that had great strength might appear with the head of a lion, or a being that was unusually smart and crafty might have some of the features of a fox. Khan theorized that this is why ancient cultures, such as the Egyptian, pictured the gods that rule the afterlife realm as having animal heads.[72]

The propensity near-death reality has for molding itself into holo-gramlike shapes that mirror the thoughts, desires, and symbols that populate our minds explains why Westerners tend to perceive the beings of light as Christian religious figures, why Indians perceive them as Hindu saints and deities, and so on. The plasticity of the ND realm suggests that such outward appearances may be no more or less real than the food wished into existence by the little girl mentioned above, the woman who appeared as an amalgam of a cobra and a rose, and the spectral clothing conjured into existence by the NDEer who was embarrassed at his own nakedness. This same plasticity explains the other cultural differences one finds in near-death experiences, such as why some NDEers reach the hereafter by traveling through a tunnel, some by crossing a bridge, some by going over a body of water, and some simply by walking down a road. Again it appears that in a reality created solely out of interacting thought structures, even the landscape itself is sculpted by the ideas and expectations of the ex-periencer.

At this juncture an important point needs to be made. As startling and foreign as the near-death realm seems, the evidence presented in this book reveals that our own level of existence may not be all that different. As we have seen, we too can access all information, it is just a little more difficult for us. We too can occasionally have personal flashforwards and come face-to-face with the phantasmal nature of time and space. And we too can sculpt and reshape our bodies, and sometimes even our reality, according to our beliefs, it just takes us a little more time and effort. Indeed, Sai Baba's abilities suggest that we can even materialize food simply by wishing for it, and Therese Neumann's inedia offers evidence that eating may ultimately be as unnecessary for us as it is for individuals in the near-death realm.

In fact, it appears that this reality and the next are different in degree, but not in kind. Both are hologramlike constructs, realities that are established, as Jahn and Dunne put it, only by the interaction of consciousness with its environment. Put another way, our reality appears to be a more frozen version of the afterlife dimension. It takes a little more time for our beliefs to resculpt our bodies into things like nail-like stigmata and for the symbolic language of our psyches to manifest externally as synchronicities. But manifest they do, in a slow and inexorable river, a river whose persistent presence teaches us that we live in a universe we are only just beginning to understand.

Information about the Near-Death Realm
from Other Sources

One does not have to be in a life-threatening crisis to visit the afterlife dimension. There is evidence that the ND realm can also be reached during OBEs. In his writings, Monroe describes several visits to levels of reality in which he encountered deceased friends.[73] An even more skilled out-of-body visitor to the land of the dead was Swedish mystic Swedenborg. Born in 1688, Swedenborg was the Leonardo da Vinci of his era. In his early years he studied science. He was the leading mathematician in Sweden, spoke nine languages, was an engraver, a politician, an astronomer, and a businessman, built watches and microscopes as a hobby, wrote books on metallurgy, color theory, commerce, economics, physics, chemistry, mining, and anatomy, and invented prototypes for the airplane and the submarine.

Throughout all of this he also meditated regularly, and when he reached middle age, developed the ability to enter deep trances during which he left his body and visited what appeared to him to be heaven and conversed with "angels" and "spirits." That Swedenborg was experiencing something profound during these journeys, there can be no doubt. He became so famous for this ability that the queen of Sweden asked him to find out why her deceased brother had neglected to respond to a letter she had sent him before his death. Swedenborg promised to consult the deceased and the next day returned with a message which the queen confessed contained information only she and her dead brother knew. Swedenborg performed this service several times for various individuals who sought his help, and on another occasion told a widow where to find a secret compartment in her deceased husband's desk in which she found some desperately needed documents. So well known was this latter incident that it inspired the German philosopher Immanuel Kant to write an entire book on Swedenborg entitled *Dreams of a Spirit-Seer*.

But the most amazing thing about Swedenborg's accounts of the afterlife realm is how closely they mirror the descriptions offered by modern-day NDEers. For example, Swedenborg talks about passing through a dark tunnel, being met by welcoming spirits, landscapes more beautiful than any on earth and one where time and space no longer exist, a dazzling light that emitted a feeling of love, appearing before beings of light, and being enveloped by an all-encompassing peace and serenity.[74] He also says that he was allowed to observe firsthand the arrival of the newly deceased in heaven, and watch as they were subjected to the life review, a process he called "the opening of the Book of Lives." He acknowledged that during the process a person witnessed "everything they had ever been or done," but added a unique twist. According to Swedenborg, the information that arose during the opening of the Book of Lives was recorded in the nervous system of the person's spiritual body. Thus, in order to evoke the life review an "angel" had to examine the individual's entire body "beginning with the fingers of each hand, and proceeding through the whole."[75]

Swedenborg also refers to the holographic thought balls the angels use to communicate and says that they are no different from the portrayals he could see in the "wave-substance" that surrounded a person. Like most NDEers he describes these telepathic bursts of knowledge as a picture language so dense with information that each image contains a thousand ideas. A communicated series of these portrayals can also be quite lengthy and "last up to several hours, in such a sequential arrangement that one can only marvel."[76]

But even here Swedenborg added a fascinating twist. In addition to using portrayals, angels also employ a speech that contains concepts that are beyond human understanding. In fact, the main reason they use portrayals is because it is the only way they can make even a pale version of their thoughts and ideas comprehensible to human beings.[77]

Swedenborg's experiences even corroborate some of the less commonly reported elements of the NDE. He noted that in the spirit world one no longer needs to eat food, but added that information takes its place as a source of nourishment.[78] He said that when spirits and angels talked, their thoughts were constantly coalescing into three-dimensional symbolic images, especially animals. For example, he said that when angels talked about love and affection "beautiful animals are presented, such as lambs. . . . When however the angels are talking about evil affections, this is portrayed by hideous, fierce, and useless animals, like tigers, bears, wolves, scorpions, snakes, and mice."[79] Although it is not a feature reported by modern NDEers, Swedenborg said that he was astonished to find that in heaven there are also spirits from other planets, an astounding assertion for a man who was born over three hundred years ago![80]

Most intriguing of all are those remarks by Swedenborg that seem to refer to reality's holographic qualities. For instance, he said that although human beings appear to be separate from one another, we are all connected in a cosmic unity. Moreover, each of us is a heaven in miniature, and every person, indeed the entire physical universe, is a microcosm of the greater divine reality. As we have seen, he also believed that underlying visible reality was a wave-substance.

In fact, several Swedenborg scholars have commented on the many parallels between some of Swedenborg's concepts and Bohm and Pribram's theory. One such scholar is Dr. George F. Dole, a professor of theology at the Swedenborg School of Religion in Newton, Massachusetts. Dole, who holds degrees from Yale, Oxford, and Harvard, notes that one of the most basic tenets of Swedenborg's thinking is that our universe is constantly created and sustained by two wavelike flows, one from heaven and one coming from our own soul or spirit. "If we put these images together, the resemblance to the hologram is striking," says Dole. "We are constituted by the intersection of two flows—one direct, from the divine, and one indirect, from the divine via our environment. We can view ourselves as interference patterns, because the inflow is a wave phenomenon, and we are where the waves meet."[81]

Swedenborg also believed that, despite its ghostlike and ephemeral qualities, heaven is actually a more fundamental level of reality than our own physical world. It is, he said, the archetypal source from which all earthly forms originate, and to which all forms return, a concept not too dissimilar from Bohm's idea of the implicate and explicate orders. In addition, he too believed that the afterlife realm and physical reality are different in degree but not in kind, and that the material world is just a frozen version of the thought-built reality of heaven. The matter that comprises both heaven and earth "flows in by stages" from the Divine, said Swedenborg, and "at each new stage it becomes more general and therefore coarser and hazier, and it becomes slower, and therefore more viscous and colder."[82]

Swedenborg filled almost twenty volumes with his experiences, and on his deathbed was asked if there was anything he wanted to recant. He earnestly replied: "Everything that I have written is as true as you now behold me. I might have said much more had it been permitted to me. After death you will see all, and then we shall have much to say to each other on the subject."[83]

The Land of Nonwhere

Swedenborg is not the only individual in history who possessed the ability to make out-of-body journeys to the subtler levels of reality. The twelfth-century Persian Sufis also employed deep trancelike meditation to visit the "land where spirits dwell." And again, the parallels between their reports and the body of evidence that has accrued in this chapter are striking. They claimed that in this other realm one possesses a "subtle body" and relies on senses that are not always associated with "specific organs" in that body. They asserted that it is a dimension populated by many spiritual teachers, or imams, and sometimes called it "the country of the hidden Imam."

They held that it is a world created solely out of the subtle matter of *alam almithal,* or thought. Even space itself, including "nearness," "distances," and "far-off" places, was created by thought. But this did not mean that the country of the hidden Imam was unreal, a world constituted out of sheer nothingness. Nor was it a landscape created by only one mind. Rather it was a plane of existence *created by the imagination of many people,* and yet one that still had its own corporeality and dimension, its own forests, mountains, and even cities. The Sufis devoted a good deal of their writings to the clarification of this point. So alien is this idea to many Western thinkers that the late Henry Corbin, a professor of Islamic Religion at the Sorbonne in Paris and a leading authority in Iranian-Islamic thought, coined the term *imaginal* to describe it, meaning a world that is created by imagination but is ontologically no less real than physical reality. "The reason I absolutely had to find another expression was that, for a good many years, my profession required me to interpret Arabic and Persian texts, whose meaning I would undoubtedly have betrayed had I simply contented myself with the term *imaginary,*" stated Corbin.[84]

Because of the imaginal nature of the afterlife realm, the Sufis concluded that *imagination itself is a faculty of perception,* an idea that offers new light on why Whitton's subject materialized a hand only after he started thinking, and why visualizing images has such a potent effect on the health and physical structure of our bodies. It also contributed to the Sufis' belief that one could use visualization, a process they called "creative prayer," to alter and reshape the very fabric of one's destiny.

In a notion that parallels Bohm's implicate and explicate orders, the Sufis believed that, despite its phantasmal qualities, the afterlife realm is the generative matrix that gives birth to the entire physical universe. All things in physical reality arise from this spiritual reality, said the Sufis. However, even the most learned among them found this strange, that by meditating and venturing deep into the psyche one arrived in an inner world that "turns out to envelop, surround, or contain that which at first was outer and visible."[85]

This realization is, of course, just another reference to the nonlocal and holographic qualities of reality. Each of us contains the whole of heaven. More than that, each of us contains the location of heaven. Or as the Sufis put it, instead of having to search for spiritual reality "in the where," the "where" is *in* us. Indeed, in discussing the nonlocal aspects of the afterlife realm, a twelfth-century Persian mystic named Sohrawardi said that the country of the hidden Imam might better be called *Na-Koja-Abad*, "the land of nonwhere."[86]

Admittedly this idea is not new. It is the same sentiment expressed in the statement "the kingdom of heaven is within." What *is* new is the idea that such notions are actually references to the nonlocal aspects of the subtler levels of reality. Again, it is suggested that when a person has an OBE they might not actually travel anywhere. They might be merely altering the always illusory hologram of reality so that they have the experience of traveling somewhere. In a holographic universe not only is consciousness already everywhere, it too is nonwhere.

The idea that the afterlife realm lies deep in the nonlocal expanse of the psyche has been alluded to by some NDEers. As one seven-year-old boy put it, "Death is like walking into your mind."[87] Bohm offers a similarly nonlocal view of what happens during our transition from this life to the next: "At the present, our whole thought process is telling us that we have to keep our attention here. You can't cross the street, for example, if you don't. But consciousness is always in the unlimited depth which is beyond space and time, in the subtler levels of the implicate order. Therefore, if you went deeply enough into the actual present, then maybe there's no difference between this moment and the next. The idea would be that in the death experience you would get into that. Contact with eternity is in the present moment, but it is mediated by thought. It is a matter of attention."[88]

Intelligent and Coordinated Images of Light

The idea that the subtler levels of reality can be accessed through a shift in consciousness alone is also one of the main premises of the yogic tradition. Many yogic practices are designed specifically to teach individuals how to make such journeys. And once again, the individuals who succeed in these ventures describe what is by now a familiar landscape. One such individual was Sri Yukteswar Giri, a little known but widely respected Hindu holy man who died in Puri, India, in 1936. Evans-Wentz, who met Sri Yukteswar in the 1920s, described him as a man of "pleasing presence and high character" fully "worthy of the veneration that his followers accorded him."[89]

Sri Yukteswar appears to have been especially gifted at passing back and forth between this world and the next and described the afterlife dimension as a world composed of "various subtle vibrations of light and color" and "hundreds of times larger than the material cosmos." He also said that it was infinitely more beautiful than our own realm of existence, and abounded with "opal lakes, bright seas, and rainbow rivers." Because it is more "vibrant with God's creative light" its weather is always pleasant, and its only climatic manifestations are occasional falls of "luminous white snow and rain of many-colored lights."

Individuals who live in this wondrous realm can materialize any body they want and can "see" with any area of their body they wish. They can also materialize any fruit or other food they desire, although they "are almost freed from any necessity of eating" and "feast only on the ambrosia of eternally new knowledge."

They communicate through a telepathic series of "light pictures," rejoice at "the immortality of friendship," realize "the indestructibility of love," feel keen pain "if any mistake is made in conduct or perception of truth," and when they are confronted with the multitude of relatives, fathers, mothers, wives, husbands, and friends acquired during their "different incarnations on earth," they are at a loss as to whom to love especially and thus learn to give "a divine and equal love to all."

What is the quintessential nature of our reality once we take up residence in this luminous land? To this question, Sri Yukteswar gave an answer that was as simple as it was holographic. In this realm where eating and even breathing are unnecessary, where a single thought can materialize a "whole garden of fragrant flowers," and all bodily injuries are "healed at once by mere willing," we are, quite simply, "intelligent and coordinated images of light."[90]

More References to Light

Sri Yukteswar is not the only yogic teacher to use such hologramlike terms when describing the subtler levels of reality. Another is Sri Aurobindo Ghose, a thinker, political activist, and mystic whom Indians revere alongside Gandhi. Born in 1872 to an upper-class Indian family, Sri Aurobindo was educated in England, where he quickly developed the reputation as a kind of prodigy. He was fluent not only in English, Hindi, Russian, German, and French, but also in ancient Sanskrit. He could read a case of books a day (as a youth he read all of the many and voluminous sacred books of India) and repeat verbatim every word on every page that he read. His powers of concentration were legendary, and it was said that he could sit studying in the same posture all night long, oblivious even to the incessant bites of the mosquitoes.

Like Gandhi, Sri Aurobindo was active in the nationalist movement in India and spent time in prison for sedition. However, despite all his intellectual and humanitarian passion, he remained an atheist until one day when he saw a wandering yogi instantaneously heal his brother of a life-threatening illness. From that point on Sri Aurobindo devoted his life to the yogic disciplines and, like Sri Yukteswar, through meditation he eventually learned to become, in his own words, "an explorer of the planes of consciousness."

It was not an easy task for Sri Aurobindo, and one of the most intractable obstacles he had to overcome to accomplish his goal was to learn how to silence the endless chatter of words and thoughts that flow unceasingly through the normal human mind. Anyone who has ever tried to empty his or her mind of *all* thought for even a moment or two knows how daunting an undertaking this is. But it is also a necessary one, for the yogic texts are quite explicit on this point. To plumb the subtler and more implicate regions of the psyche does indeed require a Bohmian shift of attention. Or as Sri Aurobindo put it, to discover the "new country within us" we must first learn how "to leave the old one behind."

It took Sri Aurobindo years to learn how to silence his mind and travel inward, but once he succeeded he discovered the same vast territory encountered by all of the other Marco Polos of the spirit that we have looked at—a realm beyond space and time, composed of a "multicolored infinity of vibrations" and peopled by nonphysical beings so far in advance of human consciousness that they make us look like children. These beings can take on any form at will, said Sri Aurobindo, the same being appearing to a Christian as a Christian saint and to an Indian as a Hindu one, although he stressed that their purpose is not to deceive, but merely to make themselves more accessible "to a particular consciousness."

According to Sri Aurobindo, in their truest form these beings appear as "pure vibration." In his two-volume work, *On Yoga,* he even likens their ability to appear as either a form or a vibration, to the wave-particle duality discovered by "modern science." Sri Aurobindo also noted that in this luminous realm one is no longer restricted to taking in information in a "point-by-point" manner, but can absorb it "in great masses," and in a single glance perceive "large extensions of space and time."

In fact, quite a number of Sri Aurobindo's assertions are indistinguishable from many of Bohm's and Pribram's conclusions. He said that most human beings possess a "mental screen" that keeps us from seeing beyond "the veil of matter," but when one learns to peer beyond this veil one finds that everything is comprised of "different intensities of luminous vibrations." He asserted that consciousness is also composed of different vibrations and believed that all matter is to some degree conscious. Like Bohm, he even asserted that psychokinesis is a direct result of the fact that all matter is to some degree conscious. If matter were not conscious, no yogi could move an object with his mind because there would be no possibility of contact between the yogi and the object, Sri Aurobindo says.

Most Bohmian of all are Sri Aurobindo's remarks about wholeness and fragmentation. According to Sri Aurobindo, one of the most important things one learns in "the great and luminous kingdoms of the Spirit," is that all separateness is an illusion, and all things are ultimately interconnected and whole. Again and again in his writings he stressed this fact, and held that it was only as one descended from the higher vibrational levels of reality to the lower that a "progressive law of fragmentation" took over. We fragment things because we exist at a lower vibration of consciousness and reality, says Sri Aurobindo, and it is our propensity for fragmentation that keeps us from experiencing the intensity of consciousness, joy, love, and delight for existence that are the norm in these higher and more subtle realms.

Just as Bohm believes that it is not possible for disorder to exist in a universe that is ultimately unbroken and whole, Sri Aurobindo believed the same was true of consciousness. If a single point of the universe were totally unconscious, the whole universe would be totally unconscious, he said, and if we perceive a pebble at the side of the road or a grain of sand under our fingernail to be lifeless and dead, our perception is again illusory and brought on only by our somnambulistic inurement with fragmentation.

Like Bohm, Sri Aurobindo's epiphanic understanding of wholeness also made him aware of the ultimate relativity of all truths and the arbitrariness of trying to divide the seamless holomovement up into "things." So convinced was he that any attempt to reduce the universe into absolute facts and unchangeable doctrine only led to distortion that he was even against religion, and all his life emphasized that the true spirituality came not from any organization or priesthood, but from the spiritual universe within:

> We must not only cut asunder the snare of the mind and the senses, but flee also from the snare of the thinker, the snare of the theologian and the church-builder, the meshes of the Word and the bondage of the Idea. All these are within us waiting to wall in the spirit with forms; but we must always go beyond, always renounce the lesser for the greater, the finite for the Infinite; we must be prepared to proceed from illumination to illumination, from experience to experience, from soul-state to soul-state. . . . Nor must we attach ourselves even to the truths we hold most securely, for they are but forms and expressions of the Ineffable who refuses to limit itself to any form or expression.[91]

But if the cosmos is ultimately ineffable, a farrago of multicolored vibrations, what are all the forms we perceive? What is physical reality? It is, said Sri Aurobindo, just "a mass of stable light."[92]

Survival in Infinity

The picture of reality reported by NDEers is remarkably self-consistent and is corroborated by the testimony of many of the world's most talented mystics as well. Even more astonishing is that as breathtaking and foreign as these subtler levels of reality are to those of us who reside in the world's more "advanced" cultures, they are mundane and familiar territories to so-called primitive peoples.

For example, Dr. E. Nandisvara Nayake Thero, an anthropologist who has lived with and studied a community of aborigines in Australia, points out that the aboriginal concept of the "dreamtime," a realm that Australian shamans visit by entering a profound trance, is almost identical to the afterlife planes of existence described in Western sources. It is the realm where human spirits go after death, and once there a shaman can converse with the dead and instantly access all knowledge. It is also a dimension in which time, space, and the other boundaries of earthly life cease to exist and one must learn to deal with infinity. Because of this, Australian shamans often refer to the afterlife as "survival in infinity."[93]

Holger Kalweit, a German ethnopsychologist with degrees in both psychology and cultural anthropology, goes Thero one better. An expert on shamanism who is also active in near-death research, Kalweit points out that virtually *all* of the world's shamanic traditions contain descriptions of this vast and extradimensional realm, replete with references to the life review, higher spiritual beings who teach and guide, food conjured up out of thought, and indescribably beautiful meadows, forests, and mountains. Indeed, not only is the ability to travel into the afterlife realm the most universal requirement for being a shaman, but NDEs are often the very catalyst that thrusts an individual into the role. For instance, the Oglala Sioux, the Seneca, the Siberian Yakut, the South American Guajiro, the Zulu, the Kenyan Kikuyu, the Korean Mu dang, the Indonesian Mentawai Islanders, and the Caribou Eskimo—all have traditions of individuals who became shamans after a life-threatening illness propelled them headlong into the afterlife realm.

However, unlike Western NDEers for whom such experiences are disorientingly new, these shamanic explorers appear to have a far vaster knowledge of the geography of these subtler realms and are often able to return to them again and again. Why? Kalweit believes it is because such experiences are a daily reality for such cultures. Whereas our society suppresses any thoughts or mention of death and dying, and has devalued the mystical by defining reality strictly in terms of the material, tribal peoples still have day-to-day contact with the psychic nature of reality. Thus, they have a better understanding of the rules that govern these inner realms, says Kalweit, and are much more skilled at navigating their territories.[94]

That these inner regions have been well traveled by shamanic peoples is evidenced by an experience anthropologist Michael Harner had among the Conibo Indians of the Peruvian Amazon. In 1960 the American Museum of Natural History sent Harner on a year-long expedition to study the Conibo, and while there he asked the Amazonian natives to tell him about their religious beliefs. They told him that if he really wished to learn, he had to take a shamanic sacred drink made from a hallucinogenic plant known as *ayahuasca*, the "soul vine." He agreed and after drinking the bitter concoction had an out-of-body experience in which he traveled a level of reality populated by what appeared to be the gods and devils of the Conibo's mythology. He saw demons with grinning crocodilian heads. He watched as an "energy-essence" rose up out of his chest and floated toward a dragon-headed ship manned by Egyptian-style figures with blue-jay heads; and he felt what he thought was the slow, advancing numbness of his own death.

But the most dramatic experience he had during his spirit journey was an encounter with a group of winged, dragonlike beings that emerged from his spine. After they had crawled out of his body, they "projected" a visual scene in front of him in which they showed him what they said was the "true" history of the earth. Through a kind of "thought language" they explained that they were responsible for both the origin and evolution of all life on the planet. Indeed, they resided not only in human beings, but in all life, and had created the multitude of living forms that populates the earth to provide themselves with a hiding place from some undisclosed enemy in outer space (Harner notes that although the beings were almost like DNA, at the time, 1961, he knew nothing of DNA).

After this concatenation of visions was over, Harner sought out a blind Conibo shaman noted for his paranormal talents to talk to him about the experience. The shaman, who had made many excursions into the spirit world, nodded occasionally as Harner related the events that had befallen him, but when he told the old man about the dragonlike beings and their claim that they were the true masters of the earth, the shaman smiled with amusement. "Oh, they're always saying that. But they are only the Masters of Outer Darkness," he corrected.

"I was stunned," says Harner. "What I had experienced was already familiar to this barefoot, blind shaman. Known to him from his

own explorations of the same hidden world into which I had ventured."
However, this was not the only shock Harner received. He also re-
counted his experience to two Christian missionaries who lived nearby,
and was intrigued when they too seemed to know what he was talking
about. After he finished they told him that some of his descriptions
were virtually identical to certain passages in the Book of Revelation,
passages that Harner, an atheist, had never read.[95] So it seems that
the old Conibo shaman perhaps was not the only individual to have
traveled the same ground Harner later and more falteringly covered.
Some of the visions and "trips to heaven" described by Old and New
Testament prophets may also have been shamanic journeys into the
inner realm.

   Is it possible that what we have been viewing as quaint folklore and
charming but naive mythology are actually sophisticated accounts of
the cartography of the subtler levels of reality? Kalweit for one believes the answer is an emphatic yes. "In light of the revolutionary
findings of recent research into the nature of dying and death, we can
no longer look upon tribal religions and their ideas about the World
of the Dead as limited conceptions," he says. "[Rather] the shaman
should be considered as a most up-to-date and knowledgeable psychologist."[96]

An Undeniable Spiritual Radiance

One last piece of evidence of the reality of the NDE is the transformative effect it has on those who experience it. Researchers have discovered that NDEers are almost always profoundly changed by their
journey to the beyond. They become happier, more optimistic, more
easygoing, and less concerned with material possessions. Most striking of all, their capacity to love expands enormously. Aloof husbands
suddenly become warm and affectionate, workaholics start relaxing
and devoting time to their families, and introverts become extroverts.
These changes are often so dramatic that people who know the NDEer
frequently remark that he or she has become an entirely different
person. There are even cases on record of criminals completely reform-
ing their ways, and fire-and-brimstone preachers replacing their message of damnation with one of unconditional love and compassion.

NDEers also become much more spiritually oriented. They return not only firmly convinced of the immortality of the human soul, but also with a deep and abiding sense that the universe is compassionate and intelligent, and this loving presence is always with them. However, this awareness does not necessarily result in their becoming more religious. Like Sri Aurobindo, many NDEers stress the importance of the distinction between religion and spirituality, and assert that it is the latter that has blossomed into greater fullness in their lives, not the former. Indeed, studies show that following their experience, NDEers display an increased openness to ideas outside their own religious background, such as reincarnation and Eastern religions.[97]

This widening of interests frequently extends to other areas as well. For instance, NDEers often develop a marked fascination for the types of subjects discussed in this book, in particular psychic phenomena and the new physics. One NDEer investigated by Ring, for example, was a driver of heavy equipment who displayed no interest in books or academic pursuits prior to his experience. However, during his NDE he had a vision of total knowledge, and although he was unable to recall the content of the vision after he recovered, various physics' terms started popping into his head. One morning not long after his experience he blurted out the word *quantum*. Later he announced cryptically, "Max Planck—you'll be hearing about him in the near future," and as time continued to pass, fragments of equations and mathematical symbols began to surface in his thoughts.

Neither he nor his wife knew what the word *quantum* meant, or who Max Planck (widely viewed as the founding father of quantum physics) was until the man went to a library and looked the words up. But after discovering that he was not talking gibberish, he started to read voraciously, not only books on physics, but also on parapsychology, metaphysics, and higher consciousness; and he even enrolled in college as a physics major. The man's wife wrote a letter to Ring trying to describe her husband's transformation:

> Many times he says a word he has never heard before in our reality—it might be a foreign word of a different language—but learns . . . it in relationship to the "light" theory. . . . He talks about things faster than the speed of light and it's hard for me to understand. . . . When [he] picks up a book on physics he already knows the answer and seems to feel more. . . .[98]

The man also started developing various psychic abilities after his experience, which is not uncommon among NDEers. In 1982 Bruce Greyson, a psychiatrist at the University of Michigan and IANDS's director of research, gave sixty-nine NDEers a questionnaire designed to study this issue, and he found that there was an increase in virtually all of the psychic and psi-related phenomena he assessed.[99] Phyllis Atwater, an Idaho housewife who became an NDE researcher following her own transformative NDE, has interviewed dozens of NDEers and has obtained similar findings. "Telepathy and healing gifts are common," she states. "So is 'remembering' the future. Time and space stop, and you live in a future sequence in detail. Then, when the event occurs, you recognize it."[100]

Moody believes that the profound and positive identity changes such individuals undergo is the most compelling evidence that NDEs are actually journeys into some spiritual level of reality. Ring agrees. "[At the core of the NDE] we find an absolute and undeniable spiritual radiance," he says. "This spiritual core is so awesome and overwhelming that the person is at once and forever thrust into an entirely new mode of being."[101]

NDE researchers are not the only individuals who are beginning to accept the existence of this dimension and the spiritual component of the human race. Nobelist Brian Josephson, himself a longtime meditator, is also convinced that there are subtler levels of reality, levels that can be accessed through meditation and where, quite possibly, one travels after death.[102]

At a 1985 symposium on the possibility of life beyond biological death held at Georgetown University and convened by U.S. Senator Claiborne Pell, physicist Paul Davies expressed a similar openness. "We are all agreed that, at least insofar as human beings are concerned, mind is a product of matter, or put more accurately, mind finds expression through matter (specifically our brains). The lesson of the quantum is that matter can only achieve concrete, well-defined existence in conjunction with mind. Clearly, if mind is *pattern* rather than *substance*, then it is capable of many different representations."[103]

Even psychoneuroimmunologist Candace Pert, another participant at the symposium, was receptive to the idea. "I think it is important to realize that information is stored in the brain, and it is conceivable to me that this information could transform itself into some other realm. Where does the information go after the destruction of the molecules (the mass) that compose it? Matter can neither be created nor destroyed, and perhaps biological information flow cannot just disappear at death and must be transformed into another realm," she says.[104]

Is it possible that what Bohm has called the implicate level of reality is actually the realm of the spirit, the source of the spiritual radiance that has transfigured the mystics of all ages? Bohm himself does not dismiss the idea. The implicate domain "could equally well be called Idealism, Spirit, Consciousness," he states with typical matter-of-factness. "The separation of the two—matter and spirit—is an abstraction. The ground is always one."[105]

Who Are the Beings of Light?

Because most of the above remarks were made by physicists and not theologians, one cannot help but wonder if perhaps the interest in new physics displayed by Ring's NDEer is an indication of something deeper. If, as Bohm suggests, physics is beginning to make inroads in areas that were once exclusively the province of the mystic, is it possible that these encroachments have already been anticipated by the beings who inhabit the near-death realm? Is that why NDEers are given an insatiable hunger for such knowledge? Are they, and by proxy the rest of the human race, being prepared for some coming confluence between science and the spiritual?

We will explore this possibility a little later. First, another question must be asked. If the existence of this higher dimension is no longer at issue, then what are its parameters? More specifically, who are the beings that inhabit it, and what is their society, dare one say their civilization, really like?

These are, of course, difficult questions to answer. When Whitton tried to find out the identity of the beings who counseled people in the between-life state, he found the answer elusive. "The impression my subjects gave—the ones who could answer the question—was that these were entities who had completed their cycle of incarnations here," he says.[106]

After hundreds of journeys into the inner realm, and after interviewing dozens of other talented fellow OBEers on the matter, Monroe has also come up empty-handed. "Whatever they may be, [these beings] have the ability to radiate a warmth of friendliness that evokes complete trust," he observes. "Perceiving our thoughts is absurdly easy for [them]." And "the entire history of humankind and earth is available to them in the most minute detail." But Monroe, too, confesses ignorance when it comes to the ultimate identity of these nonphysical entities, save that their first order of business appears to be "totally solicitous as to the well-being of the human beings with whom they are associated."[107]

Not *much* more can be said about the civilizations of these subtle realms, save that individuals who are privileged enough to visit them universally report seeing many vast and celestially beautiful cities there. NDEers, yogic adepts, and *ayahuasca*-using shamans—all describe these mysterious metropolises with remarkable consistency. The twelfth-century Sufis were so familiar with them that they even gave several of them names.

The most notable feature of these great cities is that they are brilliantly luminous. They are also frequently described as foreign in architecture, and so sublimely beautiful that, like all of the other features of these implicate dimensions, words fail to convey their grandeur. In describing one such city Swedenborg said that it was a place "of staggering architectural design, so beautiful that you would say this is the home and the source of the art itself."[108]

People who visit these cities also frequently assert that they have an unusual number of schools and other buildings associated with the pursuit of knowledge. Most of Whitton's subjects recalled spending at least some time hard at work in vast halls of learning equipped with libraries and seminar rooms while in the between-life state.[109] Many NDEers also report being shown "schools," "libraries," and "institutions of higher learning" during their experiences.[110] And one can even find references to great cities devoted to learning and reachable only by journeying into "the hidden depths of the mind" in eleventh-century Tibetan texts. Edwin Bernbaum, a Sanskrit scholar at the University of California at Berkeley, believes that James Hilton's novel *Lost Horizon*, in which he created the fictional community of Shangri-La, was actually inspired by one of these Tibetan legends.*[111]

---

*Throughout my high-school and college years I had vivid and frequent dreams that I was attending classes on spiritual subjects at a strangely beautiful university in some sublime and otherworldly place. These were not anxiety dreams about going to school, but incredibly pleasant flying dreams in which I floated weightlessly to lectures on the human energy field and reincarnation. During these dreams I sometimes encountered people I had known in this life but who had died, and even people who identified themselves as souls about to be reborn.

The only problem is that in an imaginal realm such descriptions don't mean very much. One can never be sure whether the spectacular architectural structures NDEers encounter are realities or just allegorical phantasms. For instance, both Moody and Ring have reported cases in which NDEers said that the buildings of higher learning they visited were not just devoted to knowledge, but were literally *built out of* knowledge.[112] This curious choice of words suggests that perhaps visits to these edifices are actually encounters with something so beyond human conception—perhaps a dynamic living cloud of pure knowledge, or what information becomes, as Pert puts it, after it has been *transformed into another realm*—that translating it into a hologram of a building or library is the only way the human mind can process it.

The same is true of the beings one encounters in the subtler dimensions. We can never know from their appearance alone what they really are. For example, George Russell, a well-known turn-of-the-century Irish seer and an extraordinarily talented OBEer, encountered many "beings of light" during what he called his journeys into the "inner world." When asked once during an interview to describe what these beings looked like he stated:

The first of these I saw I remember very clearly, and the manner of its appearance: there was at first a dazzle of light, and then I saw that this came from the heart of a tall figure with a body apparently shaped out of half-transparent or opalescent air, and throughout the body ran a radiant, electrical fire, to which the heart seemed the centre. Around the head of this being and through its waving luminous hair, which was blown all about the body like living strands of gold, there appeared flaming wing-like auras. From the being itself light seemed to stream outwards in every direction; and the effect left on me after the vision was one of extraordinary lightness, joyousness, or ecstasy.[113]

Intriguingly, I have met several other individuals, usually people with more than normal psychic ability, who have also had these dreams (one, a talented Texas clairvoyant named Jim Gordon, was so baffled by the experience that he often asked his nonplussed mother why he had to go to school *twice*, once during the day with all the other children, and once at night while he slept). It is relevant to mention here that Monroe and numerous other OBE researchers believe that flying dreams are actually just poorly remembered OBEs, making me wonder if perhaps some of us, at least, are visiting these incorporeal schools even while we are alive. If anyone reading this book has also had such experiences, I would be very interested in hearing about them.

On the other hand, Monroe asserts that once he has been in the presence of one of these nonphysical entities for a while, it discards its appearance and he perceives nothing, although he continues to sense "the radiation that is the entity."[114] Again the question can be asked, When a journeyer to the inner dimensions encounters a being of light, is that being a reality or just an allegorical phantasm? The answer is, of course, that it is a bit of both, for in a holographic universe *all* appearances are illusions, hologramlike images constructed by the interaction of the consciousness present, but illusions based, as Pribram says, on *something* that is there. Such are the dilemmas one faces in a universe that appears to us in explicate form but always has its source in something ineffable, in the implicate.

We can take heart in the fact that the hologramlike images our minds construct in the afterlife realm appear to bear at least some relationship to the *something* that is there. When we encounter a disembodied cloud of pure knowledge, we convert it into a school or library. When an NDEer meets a woman with whom he has had a love/hate relationship, he sees her as half rose, half cobra, a symbol that still conveys the quintessence of her character; and when travelers in the subtler realms encounter helpful, nonphysical consciousnesses, they see them as luminous and angelic beings.

As for the ultimate identity of these beings, we can deduce from their behavior that they are older, wiser, and possess some deep and loving connection to the human species, but beyond this the question remains unanswered as to whether they are gods, angels, the souls of human beings who have finished reincarnating, or something that is altogether beyond human comprehension. To speculate further would be presumptuous in that it would not only be tackling a question that thousands of years of human history have failed to resolve, but would also ignore Sri Aurobindo's warning against turning spiritual understandings into religious ones. As science gathers more evidence, the answer will most assuredly become clearer, but until then, the question of who and what these beings are remains open.

The Omnijective Universe

The hereafter is not the only realm in which we can encounter hologramlike apparitions sculptured by our beliefs. It appears that on occasion we can even have such experiences at our own level of existence. For example, philosopher Michael Grosso believes that miraculous appearances of the Virgin Mary may also be hologramlike projections created by the collective beliefs of the human race. One "Marian" vision that is especially holographic in flavor is the well-known appearance of the Virgin in Knock, Ireland, in 1879. On that occasion fourteen people saw three glowing and eerily motionless figures consisting of Mary, Joseph, and St. John the Evangelist (identified because he closely resembled a statue of the saint in a nearby village) standing in a meadow next to the local church. These brilliantly luminous figures were so real that when witnesses approached, they could even read the lettering on a book St. John was holding. But when one of the women present tried to embrace the Virgin, her arms closed on empty air. "The figures appeared so full and lifelike I could not understand why my hands could not feel what was so plain and distinct to my sight," the woman later wrote.[115]

Another impressively holographic Marian vision is the equally famous appearance of the Virgin in Zeitoun, Egypt. The sightings began in 1968 when two Moslem automobile mechanics saw a luminous apparition of Mary standing on the ledge of the central dome of a Coptic church in the poor Cairo suburb. For the next three years glowing three-dimensional images of Mary, Joseph, and the Christ Child appeared weekly over the church, sometimes hovering in midair for as long as six hours.

Unlike the figures at Knock, the Zeitoun apparitions moved about and waved at the crowds of people who regularly gathered to see them. However, they too had many holographic aspects. Their appearance was always heralded by a brilliant flash of light. Like holograms shifting from their frequency aspects and slowly coming into focus, they were at first amorphous and slowly coalesced into human shape. They were often accompanied by doves "formed of pure light" that soared for great distances over the crowd, but never flapped their wings. Most telling of all, after three years of manifestations and as interest in the phenomenon started to wane, the Zeitoun figures also waned, becoming hazier and hazier until, in their last several appearances, they were little more than clouds of luminous fog. Nonetheless, during their peak, the figures were seen by literally hundreds of thousands of witnesses and were extensively photographed. "I've interviewed quite a number of these people, and when you hear them talk about what they saw you can't get rid of the feeling that they're describing some sort of holographic projection," says Grosso.[116]

In his thought-provoking book *The Final Choice,* Grosso says that after studying the evidence he is convinced that such visions are not appearances of the historical Mary, but are actually psychic holographic projections created by the collective unconscious. Interestingly, not all of the Marian apparitions are silent. Some, like the manifestations at Fatima and Lourdes, speak, and when they do their message is invariably a warning of impending apocalypse if we mortals do not mend our ways. Grosso interprets this as evidence that the human collective unconscious is deeply disturbed by the violent impact modern science has had on human life and on the ecology of the earth. Our collective dreams are, in essence, warning us of the possibility of our own self-destruction.

Others have also agreed that belief in Mary is the motivating force that causes these projections to coalesce into being. For instance, Rogo points out that in 1925, while the Coptic church that became the site of the Zeitoun manifestations was being built, the philanthropist responsible for its construction had a dream in which the Virgin told him she would appear at the church as soon as it was completed. She did not appear at the prescribed time, but the prophecy was well known in the community. Thus *"there existed a forty-year-old tradition that a Marian visitation would eventually take place at the church,"* says Rogo. "These preoccupations may have gradually built up a psychic 'blueprint' of the Virgin within the church itself, i.e., an ever-increasing pool of psychic energy created by the thoughts of the Zeitounians which in 1968 became so high-pitched that an image of the Virgin Mary burst into physical reality!"[117] In previous writings I, too, have offered a similar explanation of Marian visions.[118]

There is evidence that some UFOs may also be some kind of hologramlike phenomenon. When people first started reporting sightings of what appeared to be spacecraft from other planets in the late 1940s, researchers who delved deeply enough into the reports to realize that at least some of them had to be taken seriously assumed that they were exactly what they appeared to be—glimpses of intelligently guided crafts from more advanced and probably extraterrestrial civilizations. However, as encounters with UFOs become more widespread—especially those involving contact with UFO occupants—and data accumulates, it becomes increasingly apparent to many researchers that these so-called spacecraft are *not* extraterrestrial in origin.

Some of the features of the phenomenon that indicate they are not extraterrestrial include the following: First, there are too many sightings; literally thousands of encounters with UFOs and their occupants have been documented, so many that it is difficult to believe they could all be actual visits from other planets. Second, UFO occupants often do not possess traits one would expect in a truly extraterrestrial lifeform; too many of them are described as humanoid beings who breathe our air, display no fear of contracting earthly viruses, are well adapted to the earth's gravity and the sun's electromagnetic emissions, display recognizable emotions in their faces, and talk our language—all of which are possible but unlikely traits in truly extraterrestrial visitors.

Third, they do not behave as extraterrestrial visitors. Instead of making the proverbial landing on the White House lawn, they appear to farmers and stranded motorists. They chase jets but don't attack. They dart around in the sky allowing dozens and sometimes hundreds of witnesses to see them, but they show no interest in making any formal contact. And often, when they contact individuals their behavior still seems illogical. For instance, one of the most commonly reported types of contact is that which involve some sort of medical examination. And yet, arguably, a civilization that possesses the technological capability to travel almost incomprehensible tracts of outer space would most assuredly possess the scientific wherewithal to obtain such information without any physical contact at all or, at the very least, without having to abduct the scores of people who appear to be legitimate victims of this mysterious phenomenon.

Finally, and most curious of all, UFOs do not even behave as physical objects do. They have been watched on radar screens to make instant ninety-degree-angle turns while traveling at enormous speeds—an antic that would rip a physical object apart. They can change size, instantly vanish into nothingness, appear out of nowhere, change color, and even change shape (traits that are also displayed by their occupants). In short, their behavior is not at all what one would expect from a physical object, but of something quite different, something with which we have become more than a little familiar in this book. As astrophysicist Dr. Jacques Vallee, one of the world's most respected UFO researchers and the model for the character LaCombe in the film *Close Encounters of the Third Kind,* stated recently, "It is the behavior of an image, or a holographic projection."[119]

As the nonphysical and hologramlike qualities of UFOs become increasingly apparent to researchers, some have concluded that rather than being from other star systems, UFOs are actually visitors from other dimensions, or levels of reality (it is important to note that not all researchers agree with this point of view, and some remain convinced that UFOs are extraterrestrial in origin). However, this explanation still does not adequately explain many of the other bizarre aspects of the phenomenon, such as why UFOs aren't making formal contact, why they behave so absurdly, and so on.

Indeed, the inadequacy of the *extra*dimensional explanation, at least in the terms in which it was initially couched, only becomes more glaring as still further unusual aspects of the UFO phenomenon come into focus. One of the more baffling of these is growing evidence that UFO encounters are less of an objective experience and more of a subjective, or psychological, one. For instance, the well-known "interrupted journey" of Betty and Barney Hill, one of the most thoroughly documented UFO abduction cases on record, seems as if it were an actual alien contact in all ways except one: the commander of the UFO was dressed in a Nazi uniform, a fact that does not make sense if the Hills' abductors were truly visitors from an alien civilization, but it does if the event was psychological in nature and more akin to a dream or hallucination, experiences that often contain obvious symbols and disconcerting flaws in logic.[120]

Other UFO encounters are even more surreal and dreamlike in character, and in the literature one can find cases in which UFO entities sing absurd songs or throw strange objects (such as potatoes) at witnesses; cases that start out as straightforward abductions aboard spacecraft but end up as hallucinogenic journeys through a series of Dantesque realities; and cases in which humanoid aliens shapeshift into birds, giant insects, and other phantasmagoric creatures.

As early as 1959, and even before much of this evidence was in, the psychological and archetypal component of the UFO phenomenon inspired Carl Jung to propose that "flying saucers" were actually a product of the collective human unconscious and a kind of modern myth in the making. In 1969, and as the mythic dimension of UFO experiences became even clearer, Vallee took the observation a step further. In his landmark book *Passport to Magonia* he points out that, far from being a new phenomenon, UFOs actually appear to be a very old phenomenon in a new guise and greatly resemble various folkloric traditions, from descriptions of elves and gnomes in European countries to medieval accounts of angels to the supernatural beings described in Native American legends.

The absurd behavior of UFO entities is the same as the mischievous behavior of elves and fairies in Celtic legends, the Norse gods, and the trickster figures among the Native Americans, says Vallee. When stripped to their underlying archetypes, all such phenomena are part of the same vast, pulsating something, a something that changes its appearance to suit the culture and time period in which it manifests, but that has been with the human race for a long, long time. What is that something? In *Passport to Magonia* Vallee provides no substantive answer and says only that it appears to be intelligent, timeless, and to be the phenomenon on which all myths are based.[121]

What, then, are UFOs and related phenomena? In *Passport to Magonia* Vallee says that we cannot rule out the possibility that they are the expression of some extraordinarily advanced nonhuman intelligence, an intelligence so beyond us that its logic appears to us only as absurdity. But if this is true, how are we to explain the conclusions of mythology experts from Mircea Eliade to Joseph Campbell that myths are an organic and necessary expression of the human race, as inevitable a human by-product as language and art? Can we really accept that the collective human psyche is so barren and jejune that it developed myths only as a response to another intelligence?

And yet, if UFOs and related phenomena are merely psychic projections, how are we to explain the physical traces they leave behind, the burnt circles and deep impressions found at the sites of landings, the unmistakable tracks they make on radar screens, and the scars and incision marks they leave on the people on whom they perform their medical examinations? In an article published in 1976, I proposed that such phenomena are difficult to categorize because we are trying to hammer them into a picture of reality that is fundamentally incorrect.[122] Given that quantum physics has shown us that mind and matter are inextricably linked, I suggested that UFOs and related phenomena are further evidence of this ultimate lack of division between the psychological and physical worlds. They are indeed a product of the collective human psyche, *but they are also quite real.* Put another way, they are something the human race has not yet learned to comprehend properly, a phenomenon that is neither subjective nor objective but "omnijective"—a term I coined to refer to this unusual state of existence (I was unaware at the time that Corbin had already coined the term *imaginal* to describe the same blurred status of reality, only in the context of the mystical experiences of the Sufis).

This point of view has become increasingly prevalent among researchers. In a recent article Ring argues that UFO encounters are imaginal experiences and are similar not only to the confrontations with the real but mind-created world individuals experience during NDEs, but also to the mythic realities shamans encounter during journeys through the subtler dimensions. They are, in short, further evidence that reality is a multilayered and mind-generated hologram.[123]

"I'm finding that I'm drawn more and more to points of view that allow me not only to acknowledge and honor the reality of these different experiences, but also to see the connections between realms that, for the most part, have been studied by different categories of scholars," states Ring. "Shamanism tends to be thrown into anthropology. UFOs tend to be thrown into whatever ufology is. NDEs are studied by parapsychologists and medical people. And Stan Grof studies psychedelic experiences from a transpersonal psychology perspective. I think there's good reason to hope that the imaginal can be, and the holographic might still prove to be, perspectives that can allow one to see not the identities, but the linkages and commonalities between these different types of experiences."[124] So convinced is Ring of the profound relationship among these at first seemingly disparate phenomena that he has recently obtained a grant to do a comparative study on people who have had UFO encounters and people who have had NDEs.

Dr. Peter M. Rojcewicz, a folklorist at the Juilliard School in New York City, has also concluded that UFOs are omnijective. In fact, he believes the time has come for folklorists to realize that probably all of the phenomena discussed by Vallee in *Passport to Magonia* are as real as they are symbolic of processes deep in the human psyche. "There exists a continuum of experiences where reality and imagination imperceptibly flow into each other," he states. Rojcewicz acknowledges that this continuum is further evidence of the Bohmian unity of all things and feels that, in light of the evidence that such phenomena are imaginal/omnijective, it is no longer defensible for folklorists to treat them simply as beliefs.[125]

Numerous other researchers, including Vallee, Grosso, and Whitley Strieber, author of the bestselling book *Communion* and one of the most famous and articulate victims of a UFO abduction, have also acknowledged the seeming omnijective nature of the phenomenon. As Strieber states, encounters with UFO beings "may be our first true quantum discovery in the large-scale world: The very act of observing it may be creating it as a concrete actuality, with sense, definition, and a consciousness of its own."[126]

In short, there is growing agreement among researchers of this mysterious phenomenon that the imaginal is not confined to the afterlife realm, but has spilled over into the seeming solidity of our sticks-and-stones world. No longer confined to the visions of shamans, the old gods have sailed their celestial barks right up to the doorstep of the computer generation, only instead of dragon-headed ships their vessels are spaceships, and they have traded in their blue-jay heads for space helmets. Perhaps we should have anticipated this spillover long ago, this merging of the Land of the Dead with our own realm, for as Orpheus, the poet-musician of Greek mythology, once warned, "The gates of Pluto must not be unlocked, within is a people of dreams."

As significant as this realization is—that the universe is not objective but omnijective, that just beyond the pale of our own safe neighborhood lies a vast otherness, a numinous landscape (more properly a mindscape) as much a part of our own psyche as it is terra incognita—it still does not shed light on the deepest mystery of all. As Carl Raschke, a faculty member in the Department of Religious Studies at the University of Denver, notes, "In the omnijective cosmos, where UFOs have their place alongside quasars and salamanders, the issue of the veridical, or hallucinatory, status of glowing, circular apparitions, becomes moot. The problem is *not* whether they exist, or in what sense they exist, but what ultimate aim they serve."[127]

In other words, what is the final identity of these beings? Again, as with entities encountered in the near-death realm, there are no clear-cut answers. On one end of the spectrum, researchers such as Ring and Grosso lean toward the idea that, despite their impingements in the world of matter, they are more psychic projection than nonhuman intelligence. Grosso, for instance, thinks that, like Marian visions, they are further evidence that the psyche of the human race is in a state of unrest. As he states, "UFOs and other extraordinary phenomena are manifestations of a disturbance in the collective unconscious of the human species."[128]

On the other end of the spectrum are those researchers who maintain that, despite their archetypal characteristics, UFOs are more alien intelligence than psychic projection. For example, Raschke believes that UFOs are "a holographic materialization from a conjugate dimension of the universe" and that this interpretation "certainly must take precedence over the psychic projection hypothesis, which flounders when one examines thoughtfully the astounding, vivid, complex, and consistent features of the 'aliens' and their 'spaceships' described by abductees."[129]

Vallee is also in this camp: "I believe that the UFO phenomenon is one of the ways through which an alien form of intelligence of incredible complexity is communicating with us *symbolically*. There is no indication that it is extraterrestrial. Instead, there is mounting evidence that it . . . [comes from] *other dimensions beyond spacetime;* from a *multiverse* which is all around us, and of which we have stubbornly refused to consider in spite of the evidence available to us for centuries."[130]

As for my own feelings, I believe that probably no single explanation can account for all of the varied aspects of the UFO phenomenon. Given the apparent vastness of the subtler levels of reality, it is easy for me to believe that there are no doubt countless nonphysical species in the higher vibratory realms. Although the abundance of UFO sightings may bode against their being extraterrestrial—given the obstacle posed by the immense interstellar distances separating the Earth from the other stars in the galaxy—in a holographic universe, a universe in which there may be an infinity of realities occupying the same space as our own world, it ceases not only to be a sticking point, but may in fact be evidence of just how unfathomably abundant with intelligent life the superhologram is.

The truth is that we simply do not have the information necessary to assess how many nonphysical species are sharing our own space. Although the physical cosmos may turn out to be an ecological Sahara, the spaceless and timeless expanses of the inner cosmos may be as rich with life as the rain forest and the coral reef. After all, research into NDEs and shamanic experiences has so far taken us only just inside the borders of this cloud-shrouded realm. We do not yet know how large its continents are or how many oceans and mountain ranges it contains.

And if we are being visited by beings who are as insubstantial and plastic in form as the bodies OBEers find themselves in after they have exteriorized, it is not at all surprising that they might appear in a chameleonlike multitude of shapes. In fact, their actual appearance may be so beyond our comprehension that it may be our own holographically organized minds that give them these shapes. Just as we convert the beings of light encountered during NDEs into religious historical personages, and clouds of pure information into libraries and institutions of learning, our minds may also be sculpting the outward appearance of the UFO phenomenon.

It is interesting to note that if this is the case, it means that the true reality of these beings is apparently so transmundane and strange that we have to plumb the deepest regions of our folk memories and mythological unconscious to find the necessary symbols to give them form. It also means that we must be exceedingly careful in interpreting their actions. For example, the medical examinations that are the centerpiece of so many UFO abductions may be only a *symbolic* representation of what is going on. Rather than probing our physical bodies, these nonphysical intelligences actually may be probing some portion of us for which we currently have no labels, perhaps the subtle anatomy of our energy selves or even our very souls. Such are the problems one faces if the phenomenon is indeed an omnijective manifestation of a nonhuman intelligence.

On the other hand, if it is possible for the faith of the citizens of Knock and Zeitoun to cause luminous images of the Virgin to coalesce into existence, for the minds of physicists to dabble around with the reality of the neutrino, and for yogis such as Sai Baba to materialize physical objects out of thin air, it only stands to reason that we would also find ourselves awash with holographic projections of our beliefs and mythologies. At least some anomalous experiences may fall into this category.

For instance, history tells us that Constantine and his soldiers saw an enormous flaming cross in the sky, a phenomenon that seems to be nothing more than a psychic exteriorization of the emotions the army responsible for nothing short of the Christianization of the pagan world was feeling on the eve of their historic undertaking. The well-known manifestation of the Angels of Mons, in which hundreds of World War I British soldiers saw an immense apparition of Saint George and a squadron of angels in the sky while fighting what was at first a losing battle at the front, in Mons, Belgium, also appears to fall into the category of psychic projection.

It is clear to me that what we are calling UFO and other folkloric experiences are really a wide range of phenomena and probably include all of the above. I have also long been of the opinion that these two explanations are not mutually exclusive. It may be that Constantine's flaming cross was also a manifestation of an extradimensional intelligence. In other words, when our collective beliefs and emotions become high-pitched enough to create a psychic projection, perhaps what we are really doing is opening a doorway between this world and the next. Perhaps the only time these intelligences can appear and interact with us is when our own potent beliefs create a kind of psychic niche for them.

Another concept from the new physics may be relevant here. After acknowledging that consciousness is the agent that allows a subatomic particle such as an electron to pop into existence, we should not therefore jump to the conclusion that we are the sole agents in this creative process, cautions University of Texas physicist John Wheeler. We are creating subatomic particles and hence the entire universe, says Wheeler, but they are also creating us. Each creates the other in what he calls a "self-reference cosmology."[131] Seen in this light, UFO entities may very well be archetypes from the collective unconscious of the human race, but we may also be archetypes in their collective unconscious. We may be as much a part of their deep psychic processes as they are of ours. Strieber has also echoed this point and says that the universe of the beings who abducted him and our own are "spinning each other together" in an act of cosmic communion.[132]

The spectrum of events we are lumping into the broad category of UFO encounters may also include phenomena with which we are not even yet familiar. For instance, researchers who believe the phenomenon is some kind of psychic projection invariably assume that it is a projection of the collective human mind. However, as we have seen in this book, in a holographic universe we can no longer view consciousness as confined solely to the brain. The fact that Carol Dryer was able to communicate with my spleen and tell me that it was upset because I had yelled at it indicates that other organs in our body also possess their own unique forms of mentality. Psychoneuroimmunologists say the same about the cells in our immune system, and according to Bohm and other physicists, even subatomic particles possess this trait. As outlandish as it sounds, some aspects of UFOs and related phenomena may be projections of these collective mentalities. Certain features of Michael Harner's encounter with the dragonlike beings certainly suggest that he was confronting a kind of visual manifestation of the intelligence of the DNA molecule. In this same vein Strieber has suggested the possibility that UFO beings are what "the force of evolution looks like when it's applied to a conscious mind."[133] We must remain open to all of these possibilities. In a universe that is conscious right down to its very depths, animals, plants, even matter itself may all be participating in the creation of these phenomena.

One thing that we do know is that in a holographic universe, a universe in which separateness ceases to exist and the innermost processes of the psyche can spill over and become as much a part of the objective landscape as the flowers and the trees, reality itself becomes little more than a mass shared dream. In the higher dimensions of existence, these dreamlike aspects become even more apparent, and indeed numerous traditions have commented on this fact. The Tibetan Book of the Dead repeatedly stresses the dreamlike nature of the afterlife realm, and this is also, of course, why the Australian aborigines refer to it as the dreamtime. Once we accept this notion, that reality at all levels is omnijective and has the same ontological status as a dream, the question becomes, Whose dream is it?

Of the religious and mythological traditions that address this question, most give the same answer, It is the dream of a single divine intelligence, of God. The Hindu Vedas and yogic texts assert again and again that the universe is God's dream. In Christianity the sentiment is summed up in the oft repeated saying, we are all thoughts in the mind of God, or as the poet Keats put it, we are all part of God's "long immortal dream."

But are we being dreamed by a single divine intelligence, by God, or are we being dreamed by the collective consciousness of all things— by all the electrons, Z particles, butterflies, neutron stars, sea cucumbers, human and nonhuman intelligences in the universe? Here again we collide headlong into the bars of our own conceptual limitations, for in a holographic universe this question is meaningless. We cannot ask if the part is creating the whole, or the whole is creating the part because *the part is the whole.* So whether we call the collective consciousness of all things "God," or simply "the consciousness of all things," it doesn't change the situation. The universe is sustained by an act of such stupendous and ineffable creativity that it simply cannot be reduced to such terms. Again it is a self-reference cosmology. Or as the Kalahari Bushmen so eloquently put it, "The dream is dreaming itself."

9

Return to the Dreamtime

Only human beings have come to a point where they no longer
know why they exist. They don't use their brains and they have
forgotten the secret knowledge of their bodies, their senses, or
their dreams. They don't use the knowledge the spirit has put
into every one of them; they are not even aware of this, and
so they stumble along blindly on the road to nowhere—a
paved highway which they themselves bulldoze and make
smooth so that they can get faster to the big empty hole which
they'll find at the end, waiting to swallow them up. It's a quick
comfortable superhighway, but I know where it leads to. I've
seen it. I've been there in my vision and it makes me shudder
to think about it.

—the Lakota shaman Lame Deer
Lame Deer Seeker of Visions

Where does the holographic model go from here? Before examining
the possible answers, we might want to see where the question has
been before. In this book I have referred to the holographic concept
as a new theory, and this is true in the sense that it is the first time
it has been presented in a scientific context. But as we have seen,
several aspects of this theory have already been foreshadowed in
various ancient traditions. They are not the only such foreshadowings,
which is intriguing, for it suggests that others have also found reason

to view the universe as holographic, or at least to intuit its holographic qualities.

For example, Bohm's idea that the universe can be viewed as the compound of two basic orders, the implicate and the explicate, can be found in many other traditions. The Tibetan Buddhists call these two aspects the void and nonvoid. The nonvoid is the reality of visible objects. The void, like the implicate order, is the birthplace of all things in the universe, which pour out of it in a "boundless flux." However, only the void is real and all forms in the objective world are illusory, existing merely because of the unceasing flux between the two orders.[1]

In turn, the void is described as "subtle," "indivisible," and "free from distinguishing characteristics." Because it is seamless totality it cannot be described in words.[2] Properly speaking, even the nonvoid cannot be described in words because it, too, is a totality in which consciousness and matter and all other things are indissoluble and whole. Herein lies a paradox, for despite its illusory nature the nonvoid still contains "an infinitely vast complex of universes." And yet its indivisible aspects are always present. As the Tibet scholar John Blofeld states, "In a universe thus composed, everything interpenetrates, and is interpenetrated by, everything else; as with the void, so with the non-void—the part *is* the whole."[3]

The Tibetans prefigured some of Pribram's thinking as well. According to Milarepa, an eleventh-century Tibetan yogin and the most renowned of the Tibetan Buddhist saints, the reason we are unable to perceive the void directly is because our unconscious mind (or, as Milarepa puts it, our "inner consciousness") is far too "conditioned" in its perceptions. This conditioning not only keeps us from seeing what he calls "the border between mind and matter," or what we would call the frequency domain, but also causes us to form a body for ourselves when we are in the between-life state and no longer have a body. "In the invisible realm of the heavens . . . the illusory mind is the great culprit," writes Milarepa, who counseled his disciples to practice "perfect seeing and contemplation" in order to realize this "Ultimate Reality."[4]

Zen Buddhists also recognize the ultimate indivisibility of reality, and indeed the main objective of Zen is to learn how to perceive this wholeness. In their book *Games Zen Masters Play,* and in words that could have been lifted right from one of Bohm's papers, Robert Sohl and Audrey Carr state, "To confuse the indivisible nature of reality with the conceptual pigeonholes of language is the basic ignorance from which Zen seeks to free us. The ultimate answers to existence are not to be found in intellectual concepts and philosophies, however sophisticated, but rather in a level of direct nonconceptual experience [of reality]."[5]

The Hindus call the implicate level of reality Brahman.[6] Brahman is formless but is the birthplace of all forms in visible reality, which appear out of it and then enfold back into it in endless flux.[7] Like Bohm, who says that the implicate order can just as easily be called spirit, the Hindus sometimes personify this level of reality and say that it is composed of pure consciousness. Thus, consciousness is not only a subtler form of matter, but it is more fundamental than matter; and in the Hindu cosmogony it is matter that has emerged from consciousness, and not the other way around. Or as the Vedas put it, the physical world is brought into being through both the "veiling" and "projecting" powers of consciousness.[8]

Because the material universe is only a second-generation reality, a creation of veiled consciousness, the Hindus say that it is transitory and unreal, or *maya*. As the Svetasvatara Upanishad states, "One should know that Nature is illusion *(maya)*, and that Brahman is the illusion maker. This whole world is pervaded with beings that are parts of him."[9] Similarly, the Kena Upanishad says that Brahman is an uncanny something "which changes its form every moment from human shape to a blade of grass."[10]

Because everything unfolds out of the irreducible totality of Brahman, the world is also a seamless whole, say the Hindus, and it is again *maya* that keeps us from realizing there is ultimately no such thing as separateness. *"Maya* severs the united consciousness so that the object is seen as other than the self and then as split up into the multitudinous objects in the universe," says the Vedic scholar Sir John Woodroffe. "And there is such objectivity as long as [humanity's] consciousness is veiled or contracted. But in the ultimate basis of experience the divergence has gone, for in it lie, in undifferentiated mass, experiencer, experience, and the experienced."[11]

This same concept can be found in Judaic thought. According to Kabbalistic tradition "the entire creation is an illusory projection of the transcendental aspects of God," says Leo Schaya, a Swiss expert on the Kabbalah. However, despite its illusory nature, it is not complete nothingness, "for every reflection of reality, even remote, broken up and transient, necessarily possesses something of its cause."[12] The idea that the creation set into motion by the God of Genesis is an illusion is reflected even in the Hebrew language, for as the Zohar, a thirteenth-century Kabbalistic commentary on the Torah and the most famous of the esoteric Judaic texts, notes, the verb *baro*, "to create," implies the idea of "creating an illusion."[13]

There are many holographic concepts in shamanistic thinking as well. The Hawaiian kahunas say that everything in the universe is infinitely interconnected and that this interconnectivity can almost be thought of as a web. The shaman, recognizing the interconnectedness of all things, sees himself at the center of this web and thus capable of affecting every other part of the universe (it is interesting to note that the concept of *maya* is also frequently likened to a web in Hindu thought).[14]

Like Bohm, who says that consciousness always has its source in the implicate, the aborigines believe that the true source of the mind is in the transcendent reality of the dreamtime. Normal people do not realize this and believe that their consciousness is in their bodies. However, shamans know this is not true, and that is why they are able to make contact with the subtler levels of reality.[15]

The Dogon people of the Sudan also believe that the physical world is the product of a deeper and more fundamental level of reality and is perpetually flowing out of and then streaming back into this more primary aspect of existence. As one Dogon elder described it, "To draw up and then return what one had drawn—that is the life of the world."[16]

In fact, the implicate/explicate idea can be found in virtually all shamanic traditions. States Douglas Sharon in his book *Wizard of the Four Winds: A Shaman's Story:* "Probably the central concept of shamanism, wherever in the world it is found, is the notion that underlying all the visible forms in the world, animate and inanimate, there exists a vital essence from which they emerge and by which they are nurtured. Ultimately everything returns to this ineffable, mysterious, impersonal unknown."[17]

The Candle and the Laser

Certainly one of the most fascinating properties of a piece of holographic film is the nonlocal way an image is distributed in its surface.

As we have seen, Bohm believes the universe itself is also organized in this manner and employs a thought experiment involving a fish and two television monitors to explain why he believes the universe is similarly nonlocal. Numerous ancient thinkers also appear to have recognized, or at least intuited, this aspect of reality. The twelfth-century Sufis summed it up by saying simply that "the macrocosm is the microcosm," a kind of earlier version of Blake's notion of seeing the world in a grain of sand.[18] The Greek philosophers Anaximenes of Miletus, Pythagoras, Heraclitus, and Plato; the ancient Gnostics; the pre-Christian Jewish philosopher Philo Judaeus; and the medieval Jewish philosopher Maimonides—all embraced the macrocosm-microcosm idea.

After a shamanic vision of the subtler levels of reality the semimythical ancient Egyptian prophet Hermes Trismegistus employed a slightly different phrasing and said that one of the main keys to knowledge was the understanding that "the without is like the within of things; the small is like the large."[19] The medieval alchemists, for whom Hermes Trismegistus became a kind of patron saint, distilled the sentiment into the motto "As above, so below." In talking about the same macrocosm-equals-microcosm idea the Hindu Visvasara Tantra uses somewhat cruder terms and states simply, "What is here is elsewhere."[20]

The Oglala Sioux medicine man Black Elk put an even more nonlocal twist on the same concept. While standing on Harney Peak in the Black Hills he witnessed a "great vision" during which he "saw more than I can tell and I understood more than I saw; for I was seeing in a sacred manner the shapes of all things in the spirit, and the shape of all shapes as they must live together as one being." One of the most profound understandings he came away with after this encounter with the ineffable was that Harney Peak was the center of the world. However, this distinction was not limited to Harney Peak, for as Black Elk put it, "Anywhere is the center of the world."[21] Over twenty-five centuries earlier the Greek philosopher Empedocles brushed up against the same sacred otherness and wrote that "God is a circle whose center is everywhere, and its circumference nowhere."[22]

Not content with mere words, some ancient thinkers resorted to even more elaborate analogies in their attempt to communicate the holographic properties of reality. To this end the author of the Hindu Avatamsaka Sutra likened the universe to a legendary network of pearls said to hang over the palace of the god Indra and "so arranged that if you look at one [pearl], you see all the others reflect in it." As the author of the Sutra explained, "In the same way, each object in the world is not merely itself, but involves every other object and, in fact, *is* everything else."[23]

Fa-Tsang, the seventh-century founder of the Hua-yen school of Buddhist thought, employed a remarkably similar analogy when trying to communicate the ultimate interconnectedness and interpenetration of all things. Fa-Tsang, who held that the whole cosmos was implicit in each of its parts (and who also believed that every point in the cosmos was its center), likened the universe to a multidimensional network of jewels, each one reflecting all others ad infinitum.[24]

When the empress Wu announced that she did not understand what Fa-Tsang meant by this image and asked him for further clarification, Fa-Tsang suspended a candle in the middle of a room full of mirrors. This, he told the empress Wu, represented the relationship of the One to the many. Then he took a polished crystal and placed it in the center of the room so that it reflected everything around it. This, he said, showed the relationship of the many to the One. However, like Bohm, who stresses that the universe is not simply a hologram but a holo-movement, Fa-Tsang stressed that his model was static and did not reflect the dynamism and constant movement of the cosmic interrelatedness among all things in the universe.[25]

In short, long before the invention of the hologram, numerous thinkers had already glimpsed the nonlocal organization of the universe and had arrived at their own unique ways to express this insight. It is worth noting that these attempts, crude as they may seem to those of us who are more technologically sophisticated, may have been far more important than we realize. For instance, it appears that the seventeenth-century German mathematician and philosopher Leibniz was familiar with the Hua-yen school of Buddhist thought. Some have argued that this was why he proposed that the universe is constituted out of fundamental entities he called "monads," each of which contains a reflection of the whole universe. What is significant is that Leibniz also gave the world integral calculus, and it was integral calculus that enabled Dennis Gabor to invent the hologram.

The Future of the Holographic Idea

And so an ancient idea, an idea that seems to find at least some expression in virtually all of the world's philosophical and metaphysical traditions, comes full circle. But if these ancient understandings can lead to the invention of the hologram, and the invention of the hologram can lead to Bohm and Pribram's formulation of the holographic model, to what new advances and discoveries might the holographic model lead? Already there are more possibilities on the horizon.

HOLOPHONIC SOUND

Drawing on Pribram's holographic model of the brain, Argentinian physiologist Hugo Zuccarelli recently developed a new recording technique that allows one to create what amounts to holograms made out of sound instead of light. Zuccarelli bases his technique on the curious fact that the human ears actually emit sound. Realizing that these naturally occurring sounds were the audio equivalent of the "reference laser" used to recreate a holographic image, he used them as the basis for a revolutionary new recording technique that reproduces sounds that are even more realistic and three-dimensional than those produced through the stereo process. He calls this new kind of sound "holophonic sound."[26]

After listening to one of Zuccarelli's holophonic recordings, a reporter for the *Times* of London wrote recently, "I stole a look at the reassuring numbers on my watch to make sure where I was. People approached from behind me where I knew there was only wall. . . . By the end of seven minutes I was getting the impression of figures, the embodiment of the voices on the tape. It is a multidimensional 'picture' created by sound."[27]

Because Zuccarelli's technique is based on the brain's own holographic way of processing sound, it appears to be as successful at fooling the ear as light holograms are at fooling the eyes. As a result, listeners often move their feet when they hear a recording of someone walking in front of them, and move their heads when they hear what sounds like a match being lit too near to their face (some reportedly even smell the match). Remarkably, because a holophonic recording has nothing to do with conventional stereophonic sound, it maintains its eerie three-dimensionality even when one listens to it through only one side of a headphone. The holographic principles involved also appear to explain why people who are deaf in one ear can still locate the source of a sound without moving their heads.

A number of major recording artists, including Paul McCartney, Peter Gabriel, and Vangelis, have approached Zuccarelli about his process, but because of patent considerations he has not yet disclosed the information necessary for a full understanding of his technique.*

UNSOLVED PUZZLES IN CHEMISTRY

Chemist Ilya Prigogine recently noted that Bohm's idea of the implicate-explicate order may help explain certain anomalous phenomena in chemistry. Science has long believed that one of the most absolute rules in the universe is that things always tend toward a greater state of disorder. If you drop a stereo off of the Empire State Building, when it crashes into the sidewalk it doesn't become more ordered and turn into a VCR. It becomes more disordered and turns into a pile of splintered parts.

Prigogine has discovered that this is not true for all things in the universe. He points out that, when mixed together, some chemicals develop into a more ordered arrangement, not a more disordered one. He calls these spontaneously appearing ordered systems "dissipative structures" and won a Nobel Prize for unraveling their mysteries. But how can a new and more complex system just suddenly pop into existence? Put another way, where do dissipative structures come from? Prigogine and others have suggested that, far from materializing out of nowhere, they are an indication of a deeper level of order in the universe, evidence of the implicate aspects of reality becoming explicate.[28]

If this is true, it could have profound implications and, among other things, lead to a deeper understanding of how new levels of complexity—such as attitudes and new patterns of behavior—pop into existence in the human consciousness and even how that most intriguing complexity of all, life itself, appeared on the earth several billion years ago.

*A sample audio cassette of holophonically recorded sound can be obtained for fifteen dollars from Interface Press, Box 42211, Los Angeles, California 90042.

NEW KINDS OF COMPUTERS

The holographic brain model has also recently been extended into the world of computers. In the past, computer scientists thought that the best way to build a better computer was simply to build a bigger computer. But in the last half decade or so, researchers have developed a new strategy, and instead of building single monolithic machines, some have started connecting scores of little computers together in "neural networks" that more closely resemble the biological structure of the human brain. Recently, Marcus S. Cohen, a computer scientist at New Mexico State University, pointed out that processors that rely on interfering waves of light passing through "multiplexed holographic gratings" might provide an even better analog of the brain's neural structure.[29] Similarly, physicist Dana Z. Anderson of the University of Colorado has recently shown how holographic gratings could be used to build an "optical memory" that exhibits associative recall.[30]

As exciting as these developments are, they are still just further refinements of the mechanistic approach to understanding the universe, advances that take place only within the material framework of reality. But as we have seen, the holographic idea's most extraordinary assertion is that the materiality of the universe may be an illusion, and physical reality may be only a small part of a vast and sentient nonphysical cosmos. If this is true, what implications does it have for the future? How do we begin to go about truly penetrating the mysteries of these subtler dimensions?

The Need for a Basic Restructuring of Science

Currently one of the best tools we have for exploring the unknown aspects of reality is science. And yet when it comes to explaining the psychic and spiritual dimensions of human existence, science in the main has repeatedly fallen short of the mark. Clearly, if science is to advance further in these areas, it needs to undergo a basic restructuring, but what specifically might such a restructuring entail?

Obviously the first and most necessary step is to accept the existence of psychic and spiritual phenomena. Willis Harman, the president of the Institute of Noetic Sciences and a former senior social scientist at Stanford Research Institute International, feels this acceptance is crucial not only to science, but to the survival of human civilization. Moreover, Harman, who has written extensively on the need for a basic restructuring of science, is astonished that this acceptance has not yet taken place. "Why don't we assume that any class of experiences or phenomena that have been reported, through the ages and across cultures, has a face validity that cannot be denied?" he asks.[31]

As has been mentioned, at least part of the reason is the longstanding bias Western science has against such phenomena, but the issue is not quite so simple as this. Consider for example the past-life memories of people under hypnosis. Whether these are actual memories of previous lives or not has yet to be proved, but the fact remains, the human unconscious has a natural propensity for generating at least *apparent* memories of previous incarnations. In general, the orthodox psychiatric community ignores this fact. Why?

At first glance the answer would appear to be because most psychiatrists just don't believe in such things, but this is not necessarily the case. Florida psychiatrist Brian L. Weiss, a graduate of the Yale School of Medicine and currently chairman of psychiatry at Mount Sinai Medical Center in Miami, says that since the publication of his best-selling book *Many Lives, Many Masters* in 1988—in which he discusses how he turned from being a skeptic to a believer in reincarnation after one of his patients started talking spontaneously about her past lives while under hypnosis—he has been deluged with letters and telephone calls from psychiatrists who say that they, too, are secret believers. "I think that is just the tip of the iceberg," says Weiss. "There are psychiatrists who write me they've been doing regression therapy for ten to twenty years, in the privacy of their office, and 'please don't tell anyone, but. . . .' Many *are* receptive to it, but they won't admit it."[32]

Similarly, in a recent conversation with Whitton when I asked him if he felt reincarnation would ever become an accepted scientific fact, he replied, "I think it already is. My experience with scientists is that if they've read the literature, they believe in reincarnation. The evidence is just so compelling that intellectual assent is virtually natural."[33]

Weiss's and Whitton's opinions seem borne out by a recent survey on psychic phenomena. After being assured that their replies would remain anonymous, 58 percent of the 228 psychiatrists who responded (many of them the heads of departments and the deans of medical schools) said that they believed "an understanding of psychic phenomena" was important to future graduates of psychiatry! Forty-four percent admitted believing that psychic factors were important in the healing process.[34]

So it appears that fear of ridicule may be as much if not more of a stumbling block as disbelief in getting the scientific establishment to begin to treat psychic research with the seriousness it deserves. We need more trailblazers like Weiss and Whitton (and the myriad other courageous researchers whose work has been discussed in this book) to go public with their private beliefs and discoveries. In brief, we need the parapsychological equivalent of a Rosa Parks.

Another feature that must be a part of the restructuring of science is a broadening of the definition of what constitutes scientific evidence. Psychic and spiritual phenomena have played a significant role in human history and have helped shape some of the most fundamental aspects of our culture. But because they are not easy to rope in and scrutinize in a laboratory setting, science has tended to ignore them.

Even worse, when they are studied, it is often the least important aspects of the phenomena that are isolated and catalogued. For instance, one of the few discoveries regarding OBEs that is considered valid in a scientific sense is that the brain waves change when an OBEer exits the body. And yet, when one reads accounts like Monroe's, one realizes that if his experiences are real, they involve discoveries that could arguably have as much impact on human history as Columbus's discovery of the New World or the invention of the atomic bomb. Indeed, those who have watched a truly talented clairvoyant at work know immediately that they have witnessed something far more profound than is conveyed in the dry statistics of R. H. and Louisa Rhine.

This is not to say that the Rhines' work is not important. But when vast numbers of people start reporting the same experiences, their anecdotal accounts should also be viewed as important evidence. They should not be dismissed merely because they cannot be documented as rigorously as other and often less significant features of the same phenomenon can be documented. As Stevenson states, "I believe it is better to learn what is probable about important matters than to be certain about trivial ones."[35]

It is worth noting that this rule of thumb is already applied to other more accepted natural phenomena. The idea that the universe began in a single, primordial explosion, or Big Bang, is accepted without question by most scientists. And this is odd because, although there are compelling reasons to believe that this is true, no one has ever proved that it is true. On the other hand, if a near-death psychologist were to state flatly that the realm of light NDEers travel to during their experiences is an actual other level of reality, the psychologist would be attacked for making a statement that cannot be proved. And this is odd, for there are equally compelling reasons to believe this is true. In other words, science already accepts what is probable about very important matters *if* those matters fall into the category of "fashionable things to believe," but not if they fall into the category of "unfashionable things to believe." This double standard must be eliminated before science can begin to make significant inroads into the study of both psychic and spiritual phenomena.

Most crucial of all, science must replace its enamorment with objectivity—the idea that the best way to study nature is to be detached, analytical, and dispassionately objective—with a more participatory approach. The importance of this shift has been stressed by numerous researchers, including Harman. We have also seen evidence of its necessity repeatedly throughout this book. In a universe in which the consciousness of a physicist affects the reality of a subatomic particle, the attitude of a doctor affects whether or not a placebo works, the mind of an experimenter affects the way a machine operates, and the imaginal can spill over into physical reality, we can no longer pretend that we are separate from that which we are studying. In a holographic and omnijective universe, a universe in which all things are part of a seamless continuum, strict objectivity ceases to be possible.

This is especially true when studying psychic and spiritual phenomena and appears to be why some laboratories are able to achieve spectacular results when performing remote-viewing experiments, and some fail miserably. Indeed, some researchers in the paranormal field have already shifted from a strictly objective approach to a more participatory approach. For example, Valerie Hunt discovered that her experimental results were affected by the presence of individuals who had been drinking alcohol and thus won't allow any such individuals in her lab while she is taking measurements. In this same vein, Russian parapsychologists Dubrov and Pushkin have found that they have more success duplicating the findings of other parapsychologists if they hypnotize all of the test subjects present. It appears that hypnosis eliminates the interference caused by the conscious thoughts and beliefs of the test subjects, and helps produce "cleaner" results.[36] Although such practices may seem odd in the extreme to us today, they may become standard operating procedures as science unravels further secrets of the holographic universe.

A shift from objectivity to participation will also most assuredly affect the role of the scientist. As it becomes increasingly apparent that it is the *experience* of observing that is important, and not just the act of observation, it is logical to assume that scientists in turn will see themselves less and less as observers and more and more as experiencers. As Harman states, "A willingness to be transformed is an essential characteristic of the participatory scientist."[37]

Again, there is evidence that a few such transformations are already taking place. For instance, instead of just observing what happened to the Conibo after they consumed the soul-vine *ayahuasca*, Harner imbibed the hallucinogen himself. It is obvious that not all anthropologists would be willing to take such a risk, but it is also clear that by becoming a participant instead of just an observer, he was able to learn much more than he ever could have by just sitting on the sidelines and taking notes.

Harner's success suggests that instead of just interviewing NDEers, OBEers, and other journeyers into the subtler realms, participatory scientists of the future may devise methods of traveling there themselves. Already lucid-dream researchers are exploring and reporting back on their own lucid-dream experiences. Others may develop different and even more novel techniques for exploring the inner dimensions. For instance, although not a scientist in the strictest definition of the term, Monroe has developed recordings of special rhythmic sounds that he feels facilitate out-of-body experiences. He has also founded a research center called the Monroe Institute of Applied Sciences in the Blue Ridge Mountains and claims to have trained hundreds of individuals to make the same out-of-body journeys he has made. Are such developments harbingers of the future, foreshadowings of a time when not only astronauts but "psychonauts" become the heroes we watch on the evening news?

An Evolutionary Thrust toward Higher Consciousness

Science may not be the only force that offers us passage to the land of nowhere. In his book *Heading toward Omega* Ring points out that there is compelling evidence that NDEs are on the increase. As we have seen, in tribal cultures individuals who have NDEs are often so transformed that they become shamans. Modern NDEers become spiritually transformed as well, mutating from their pre-NDE personalities into more loving, compassionate, and even more psychic individuals. From this Ring concludes that perhaps what we are witnessing is "the *shamanizing of modern humanity.*"[38] But if this is so, why are NDEs increasing? Ring believes that the answer is as simple as it is profound; what we are witnessing is "*an evolutionary thrust toward higher consciousness for all humanity.*"

And NDEs may not be the only transformative phenomenon bubbling up from the collective human psyche. Grosso believes that the increase in Marian visions during the last century has evolutionary implications as well. Similarly, numerous researchers, including Raschke and Vallee, feel that the explosion of UFO sightings in the last several decades has evolutionary significance. Several investigators, including Ring, have pointed out that UFO encounters actually resemble shamanic initiations and may be further evidence of the shamanizing of modern humanity. Strieber agrees. "I think it's rather obvious that, whether [the UFO phenomenon is being] done by somebody or [is happening] naturally, what we're dealing with is an exponential leap from one species to another. I would suspect that what we're looking at is the process of evolution in action."[39]

If such speculations are true, what is the purpose of this evolutionary transformation? There appears to be two answers. Numerous ancient traditions speak of a time when the hologram of physical reality was much more plastic than it is now, much more like the amorphous and fluid reality of the afterlife dimension. For example, the Australian aborigines say that there was a time when the entire world was dreamtime. Edgar Cayce echoed this sentiment and asserted that the earth was "at first merely in the nature of thought-forms or visualization made by pushing themselves out of themselves in whatever manner desired. . . . Then came materiality as such into the earth, through Spirit pushing itself into matter."[40]

The aborigines assert that the day will come when the earth returns to the dreamtime. In the spirit of pure speculation, one might wonder if, as we learn to manipulate the hologram of reality more and more, we will see the fulfillment of this prophecy. As we become more adept at tinkering with what Jahn and Dunne call the interface between consciousness and its environment, is it possible for us to experience a reality that is once again malleable? If this is true, we will need to learn much more than we presently know to manipulate such a plastic environment safely, and perhaps that is one purpose of the evolutionary processes that seem to be unfolding in our midst.

Many ancient traditions also assert that humanity did not originate on the earth, and that our true home is with God, or at least in a nonphysical and more paradisiacal realm of pure spirit. For instance, there is a Hindu myth that human consciousness began as a ripple that decided to leave the ocean of "consciousness as such, timeless, spaceless, infinite and eternal."[41] Awakening to itself, it forgot that it was a part of this infinite ocean, and felt isolated and separated. Loye has argued that Adam and Eve's expulsion from the Garden of Eden may also be a version of this myth, an ancient memory of how human consciousness, somewhere in its unfathomable past, left its home in the implicate and forgot that it was a part of the cosmic wholeness of all things.[42] In this view the earth is a kind of playground "in which one is free to experience all the pleasures of the flesh provided one realizes that one is a holographic projection of a . . . higher-order spatial dimension."[43]

If this is true, the evolutionary fires that are beginning to flicker and dance through our collective psyche may be our wake-up call, the trumpet note informing us that our true home is elsewhere and we can return there if we wish. Strieber, for one, believes this is precisely why UFOs are here: "I think that they are probably midwifing our birth into the nonphysical world—which is their origin. My impression is that the physical world is only a small instant in a much larger context and that reality is primarily unfolding in a non-physical way. I don't think that physical reality is the original source of being. I think that being, as consciousness, probably predates the physical."[44]

Writer Terence McKenna, another longtime supporter of the holographic model, agrees:

What this seems to be about is that from the time of the awareness of the existence of the soul until the resolution of the apocalyptic potential, there are roughly fifty thousand years. We are now, there can be no doubt, in the final historical seconds of that crisis—a crisis which involves the end of history, our departure from the planet, [and] the triumph over death. We are, in fact, closing distance with the most profound event a planetary ecology can encounter—the freeing of life from the dark chrysalis of matter.[45]

Of course these are only speculations. But whether we are on the very brink of a transition, as Strieber and McKenna suggest, or whether that watershed is still some ways off in the future, it is apparent that we are following some track of spiritual evolution. Given the holographic nature of the universe, it is also apparent that at least something like the above two possibilities awaits us somewhere and somewhen.

And lest we be tempted to assume that freedom from the physical is the end of human evolution, there is evidence that the more plastic and imaginal realm of the hereafter is also a mere stepping stone. For example, Swedenborg said that beyond the heaven he visited was another heaven, one so brilliant and formless to his perceptions that it appeared only as "a streaming of light."[46] NDEers have also occasionally described these even more unfathomably tenuous realms. "There are many higher planes, and to get back to God, to reach the plane where His spirit resides, you have to drop your garment each time until your spirit is truly free," states one of Whitton's subjects. "The learning process never stops. . . . Sometimes we are allowed glimpses of the higher planes—each one is lighter and brighter than the one before."[47]

It may be frightening to some that reality seems to become increasingly frequency-like as one penetrates deeper into the implicate. And this is understandable. It is obvious that we are still like children who need the security of a coloring book, not yet ready to draw free-form and without lines to guide our clumsy hands. To be plunged into Swedenborg's realm of streaming light would be tantamount to plunging us into a completely fluid LSD hallucination. And we are not yet mature enough or in enough control of our emotions, attitudes, and beliefs to deal with the monsters our psyches would create for ourselves there.

But perhaps that is why we are learning how to deal with small doses of the omnijective here, in the form of the relatively limited confrontations with the imaginal that UFOs and other similar experiences provide.

And perhaps that is why the beings of light tell us again and again that the purpose of life is to learn.

We are indeed on a shaman's journey, mere children struggling to become technicians of the sacred. We are learning how to deal with the plasticity that is part and parcel of a universe in which mind and reality are a continuum, and in this journey one lesson stands out above all others. As long as the formlessness and breathtaking freedom of the beyond remain frightening to us, we will continue to dream a hologram for ourselves that is comfortably solid and well defined.

But we must always heed Bohm's warning that the conceptual pigeonholes we use to parse out the universe are of our own making. They do not exist "out there," for "out there" is only the indivisible totality. Brahman. And when we outgrow any given set of conceptual pigeonholes we must always be prepared to move on, to advance from soul-state to soul-state, as Sri Aurobindo put it, and from illumination to illumination. For our purpose appears to be as simple as it is endless.

We are, as the aborigines say, just learning how to survive in infinity.

Notes

INTRODUCTION

1. Irvin L. Child, "Psychology and Anomalous Observations," *American Psychologist* 40, no. 11 (November 1985), pp. 1219–30.

1. THE BRAIN AS HOLOGRAM

1. Wilder Penfield, *The Mystery of the Mind: A Critical Study of Consciousness and the Human Brain* (Princeton, N.J.: Princeton University Press, 1975).
2. Karl Lashley, "In Search of the Engram," in *Physiological Mechanisms in Animal Behavior* (New York: Academic Press, 1950), pp. 454–82.
3. Karl Pribram, "The Neurophysiology of Remembering," *Scientific American* 220 (January 1969), p. 75.
4. Karl Pribram, *Languages of the Brain* (Monterey, Calif.: Wadsworth Publishing, 1977), p. 123.
5. Daniel Goleman, "Holographic Memory: Karl Pribram Interviewed by Daniel Goleman," *Psychology Today* 12, no. 9 (February 1979), p. 72.
6. J. Collier, C. B. Burckhardt, and L. H. Lin, *Optical Holography* (New York: Academic Press, 1971).
7. Pieter van Heerden, "Models for the Brain," *Nature* 227 (July 25, 1970), pp. 410–11.
8. Paul Pietsch, *Shufflebrain: The Quest for the Hologramic Mind* (Boston: Houghton Mifflin, 1981), p. 78.
9. Daniel A. Pollen and Michael C. Tractenberg, "Alpha Rhythm and Eye Movements in Eidetic Imagery," *Nature* 237 (May 12, 1972), p. 109.
10. Pribram, *Languages*, p. 169.

11. Paul Pietsch, "Shufflebrain," *Harper's Magazine* 244 (May 1972), p. 66.
12. Karen K. DeValois, Russell L. DeValois, and W. W. Yund, "Responses of Striate Cortex Cells to Grating and Checkerboard Patterns," *Journal of Physiology*, vol. 291 (1979), pp. 483–505.
13. Goleman, *Psychology Today*, p. 71.
14. Larry Dossey, *Space, Time, and Medicine* (Boston: New Science Library, 1982), pp. 108–9.
15. Richard Restak, "Brain Power: A New Theory," *Science Digest* (March 1981), p. 19.
16. Richard Restak, *The Brain* (New York: Warner Books, 1979), p. 253.

2. THE COSMOS AS HOLOGRAM

1. Basil J. Hiley and F. David Peat, "The Development of David Bohm's Ideas from the Plasma to the Implicate Order," in *Quantum Implications*, ed. Basil J. Hiley and F. David Peat (London: Routledge & Kegan Paul, 1987), p. 1.
2. Nick Herbert, "How Large is Starlight? A Brief Look at Quantum Reality," *Revision* 10, no. 1 (Summer 1987), pp. 31–35.
3. Albert Einstein, Boris Podolsky, and Nathan Rosen, "Can Quantum-Mechanical Description of Physical Reality Be Considered Complete?" *Physical Review* 47 (1935), p. 777.
4. Hiley and Peat, *Quantum*, p. 3.
5. John P. Briggs and F. David Peat, *Looking Glass Universe* (New York: Simon & Schuster, 1984), p. 96.
6. David Bohm, "Hidden Variables and the Implicate Order," in *Quantum Implications*, ed. Basil J. Hiley and F. David Peat (London: Routledge & Kegan Paul, 1987), p. 38.
7. "Nonlocality in Physics and Psychology: An Interview with John Stewart Bell," *Psychological Perspectives* (Fall–Winter 1988), p. 306.
8. Robert Temple, "An Interview with David Bohm," *New Scientist* (November 11, 1982), p. 362.
9. Bohm, *Quantum*, p. 40.
10. David Bohm, *Wholeness and the Implicate Order* (London: Routledge & Kegan Paul, 1980), p. 205.
11. Private communication with author, October 28, 1988.
12. Bohm, *Wholeness*, p. 192.
13. Paul Davies, *Superforce* (New York: Simon & Schuster, 1984), p. 48.
14. Lee Smolin, "What is Quantum Mechanics Really About?" *New Scientist* (October 24, 1985), p. 43.
15. Private communication with author, October 14, 1988.

16. Saybrook Publishing Company, *The Reach of the Mind: Nobel Prize Conversations* (Dallas, Texas: Saybrook Publishing Co., 1985), p. 91.

17. Judith Hooper, "An Interview with Karl Pribram," *Omni* (October 1982), p. 135.

18. Private communication with author, February 8, 1989.

19. Renee Weber, "The Enfolding-Unfolding Universe: A Conversation with David Bohm," in *The Holographic Paradigm*, ed. Ken Wilber (Boulder, Colo.: New Science Library, 1982), pp. 83–84.

20. Ibid., p. 73.

3. THE HOLOGRAPHIC MODEL AND PSYCHOLOGY

1. Renee Weber, "The Enfolding-Unfolding Universe: A Conversation with David Bohm," in *The Holographic Paradigm*, ed. Ken Wilber (Boulder, Colo.: New Science Library, 1982), p. 72.

2. Robert M. Anderson, Jr., "A Holographic Model of Transpersonal Consciousness," *Journal of Transpersonal Psychology* 9, no. 2 (1977), p. 126.

3. Jon Tolaas and Montague Ullman, "Extrasensory Communication and Dreams," in *Handbook of Dreams*, ed. Benjamin B. Wolman (New York: Van Nostrand Reinhold, 1979), pp. 178–79.

4. Private communication with author, October 31, 1988.

5. Montague Ullman, "Wholeness and Dreaming," in *Quantum Implications*, ed. Basil J. Hiley and F. David Peat (New York: Routledge & Kegan Paul, 1987), p. 393.

6. I. Matte-Blanco, "A Study of Schizophrenic Thinking: Its Expression in Terms of Symbolic Logic and Its Representation in Terms of Multidimensional Space," *International Journal of Psychiatry* 1, no. 1 (January 1965), p. 93.

7. Montague Ullman, "Psi and Psychopathology," paper delivered at the American Society for Psychical Research conference on Psychic Factors in Psychotherapy, November 8, 1986.

8. See Stephen LaBerge, *Lucid Dreaming* (Los Angeles: Jeremy P. Tarcher, 1985).

9. Fred Alan Wolf, *Star Wave* (New York: Macmillan, 1984), p. 238.

10. Jayne Gackenbach, "Interview with Physicist Fred Alan Wolf on the Physics of Lucid Dreaming," *Lucidity Letter* 6, no. 1 (June 1987), p. 52.

11. Fred Alan Wolf, "The Physics of Dream Consciousness: Is the Lucid Dream a Parallel Universe?" *Second Lucid Dreaming Symposium Proceedings/Lucidity Letter* 6, no. 2 (December 1987), p. 133.

12. Stanislav Grof, *Realms of the Human Unconscious* (New York: E. P. Dutton, 1976), p. 20.

13. Ibid., p. 236.
14. Ibid., pp. 159–60.
15. Stanislav Grof, *The Adventure of Self-Discovery* (Albany, N.Y.: State University of New York Press, 1988), pp. 108–9.
16. Stanislav Grof, *Beyond the Brain* (Albany, N.Y.: State University of New York Press, 1985), p. 31.
17. Ibid., p. 78.
18. Ibid., p. 89.
19. Edgar A. Levenson, "A Holographic Model of Psychoanalytic Change," *Contemporary Psychoanalysis* 12, no. 1 (1975), p. 13.
20. Ibid., p. 19.
21. David Shainberg, "Vortices of Thought in the Implicate Order," in *Quantum Implications*, ed. Basil J. Hiley and F. David Peat (New York: Routledge & Kegan Paul, 1987), p. 402.
22. Ibid., p. 411.
23. Frank Putnam, *Diagnosis and Treatment of Multiple Personality Disorder* (New York: Guilford, 1988), p. 68.
24. "Science and Synchronicity: A Conversation with C. A. Meier," *Psychological Perspectives* 19, no. 2 (Fall–Winter 1988), p. 324.
25. Paul Davies, *The Cosmic Blueprint* (New York: Simon & Schuster, 1988), p. 162
26. F. David Peat, *Synchronicity: The Bridge between Mind and Matter* (New York: Bantam Books, 1987), p. 235.
27. Ibid., p. 239.

4. I SING THE BODY HOLOGRAPHIC

1. Stephanie Matthews-Simonton, O. Carl Simonton, and James L. Creighton, *Getting Well Again* (New York: Bantam Books, 1980), pp. 6–12.
2. Jeanne Achterberg, "Mind and Medicine: The Role of Imagery in Healing," *ASPR Newsletter* 14, no. 3 (June 1988), p. 20.
3. Jeanne Achterberg, *Imagery in Healing* (Boston, Mass.: New Science Library, 1985), p. 134.
4. Private communication with author, October 28, 1988.
5. Achterberg, *ASPR Newsletter*, p. 20.
6. Achterberg, *Imagery*, pp. 78–79.
7. Jeanne Achterberg, Ira Collerain, and Pat Craig, "A Possible Relationship between Cancer, Mental Retardation, and Mental Disorders," *Journal of Social Science and Medicine* 12 (May 1978), pp. 135–39.
8. Bernie S. Siegel, *Love, Medicine, and Miracles* (New York: Harper & Row, 1986), p. 32.

9. Achterberg, *Imagery*, pp. 182–87.

10. Bernie S. Siegel, *Love*, p. 29.

11. Charles A. Garfield, *Peak Performance: Mental Training Techniques of the World's Greatest Athletes* (New York: Warner Books, 1984), p. 16.

12. Ibid., p. 62.

13. Mary Orser and Richard Zarro, *Changing Your Destiny* (New York: Harper & Row, 1989), p. 60.

14. Barbara Brown, *Supermind: The Ultimate Energy* (New York: Harper & Row, 1980), p. 274: as quoted in Larry Dossey, *Space, Time, and Medicine* (Boston, Mass.: New Science Library, 1982), p. 112.

15. Brown, *Supermind*, p. 275: as quoted in Dossey, *Space*, pp. 112–13.

16. Larry Dossey, *Space, Time, and Medicine* (Boston, Mass.: New Science Library, 1982), p. 112.

17. Private communication with author, February 8, 1989.

18. Brendan O'Regan, "Healing, Remission, and Miracle Cures," *Institute of Noetic Sciences Special Report* (May 1987), p. 3.

19. Lewis Thomas, *The Medusa and the Snail* (New York: Bantam Books, 1980), p. 63.

20. Thomas J. Hurley III, "Placebo Effects: Unmapped Territory of Mind/Body Interactions," *Investigations* 2, no. 1 (1985), p. 9.

21. Ibid.

22. Steven Locke and Douglas Colligan, *The Healer Within* (New York: New American Library, 1986), p. 224.

23. Ibid., p. 227.

24. Bruno Klopfer, "Psychological Variables in Human Cancer," *Journal of Prospective Techniques* 31 (1957), pp. 331–40.

25. O'Regan, *Special Report*, p. 4.

26. G. Timothy Johnson and Stephen E. Goldfinger, *The Harvard Medical School Health Letter Book* (Cambridge, Massachusetts: Harvard University Press, 1981), p. 416.

27. Herbert Benson and David P. McCallie, Jr., "Angina Pectoris and the Placebo Effect," *New England Journal of Medicine* 300, no. 25 (1979), pp. 1424–29.

28. Johnson and Goldfinger, *Health Letter Book*, p. 418.

29. Hurley, *Investigations*, p. 10.

30. Richard Alpert, *Be Here Now* (San Cristobal, N.M.: Lama Foundation, 1971).

31. Lyall Watson, *Beyond Supernature* (New York: Bantam Books, 1988), p. 215.

32. Ira L. Mintz, "A Note on the Addictive Personality," *American Journal of Psychiatry* 134, no. 3 (1977), p. 327.

33. Alfred Stelter, *Psi-Healing* (New York: Bantam Books, 1976), p. 8.

34. Thomas J. Hurley III, "Placebo Learning: The Placebo Effect as a Conditioned Response," *Investigations* 2, no. 1 (1985), p. 23.

35. O'Regan, *Special Report*, p. 3.

36. As quoted in Thomas J. Hurley III, "Varieties of Placebo Experience: Can One Definition Encompass Them All?" *Investigations* 2, no. 1 (1985), p. 13.

37. Daniel Seligman, "Great Moments in Medical Research," *Fortune* 117, no. 5 (February 29, 1988), p. 25.

38. Daniel Goleman, "Probing the Enigma of Multiple Personality," *New York Times* (June 25, 1988), p. C1.

39. Private communication with author, January 11, 1990.

40. Richard Restak, "People with Multiple Minds," *Science Digest* 92, no. 6 (June 1984), p. 76.

41. Daniel Goleman, "New Focus on Multiple Personality," *New York Times* (May 21, 1985), p. C1.

42. Truddi Chase, *When Rabbit Howls* (New York: E. P. Dutton, 1987), p. x.

43. Thomas J. Hurley III, "Inner Faces of Multiplicity," *Investigations* 1, no. 3/4 (1985), p. 4.

44. Thomas J. Hurley III, "Multiplicity & the Mind-Body Problem: New Windows to Natural Plasticity," *Investigations* 1, no. 3/4 (1985), p. 19.

45. Bronislaw Malinowski, "Baloma: The Spirits of the Dead in the Trobriand Islands," *Journal of the Royal Anthropological Institute of Great Britain and Ireland* 46 (1916), pp. 353–430.

46. Watson, *Beyond Supernature*, pp. 58–60.

47. Joseph Chilton Pearce, *The Crack in the Cosmic Egg* (New York: Pocket Books, 1974), p. 86.

48. Pamela Weintraub, "Preschool?" *Omni* 11, no. 11 (August 1989), p. 38.

49. Kathy A. Fackelmann, "Hostility Boosts Risk of Heart Trouble," *Science News* 135, no. 4 (January 28, 1989), p. 60.

50. Steven Locke, in *Longevity* (November 1988), as quoted in "Your Mind's Healing Powers," *Reader's Digest* (September 1989), p. 5.

51. Bruce Bower, "Emotion-Immunity Link in HIV Infection," *Science News* 134, no. 8 (August 20, 1988), p. 116.

52. Donald Robinson, "Your Attitude Can Make You Well," *Reader's Digest* (April 1987), p. 75.

53. Daniel Goleman in the *New York Times* (April 20, 1989), as quoted in "Your Mind's Healing Powers," *Reader's Digest* (September 1989), p. 6.

54. Robinson, *Reader's Digest*, p. 75.

55. Signe Hammer, "The Mind as Healer," *Science Digest* 92, no. 4 (April 1984), p. 100.

56. John Raymond, "Jack Schwarz: The Mind Over Body Man," *New Realities* 11, no. 1 (April 1978), pp. 72–76; see also, "Jack Schwarz: Probing ... but No Needles Anymore," *Brain/Mind Bulletin* 4, no. 2 (December 4, 1978), p. 2.

57. Stelter, *Psi-Healing,* pp. 121–24.

58. Donna and Gilbert Grosvenor, "Ceylon," *National Geographic* 129, no. 4 (April 1966).

59. D. D. Kosambi, "Living Prehistory in India," *Scientific American* 216, no. 2 (February 1967), p. 104.

60. A. A. Mason, "A Case of Congenital Ichthyosiform," *British Medical Journal* 2 (1952), pp. 422–23.

61. O'Regan, *Special Report,* p. 9.

62. D. Scott Rogo, *Miracles* (New York: Dial Press, 1982), p. 74.

63. Herbert Thurston, *The Physical Phenomena of Mysticism* (Chicago: Henry Regnery Company, 1952), pp. 120–29.

64. Thomas of Celano, *Vita Prima* (1229), as quoted by Thurston, *Physical Phenomena,* pp. 45–46.

65. Alexander P. Dubrov and Veniamin N. Pushkin, *Parapsychology and Contemporary Science,* trans. Aleksandr Petrovich (New York: Plenum, 1982), p. 50.

66. Thurston, *Physical Phenomena,* p. 68.

67. Ibid.

68. Charles Fort, *The Complete Books of Charles Fort* (New York: Dover, 1974), p. 1022.

69. Ibid., p. 964.

70. Private communication with author, November 3, 1988.

71. Candace Pert with Harris Dienstfrey, "The Neuropeptide Network," in *Neuroimmunomodulation: Interventions in Aging and Cancer,* ed. Walter Pierpaoli and Novera Herbert Spector (New York: New York Academy of Sciences, 1988), pp. 189–94.

72. Terrence D. Oleson, Richard J. Kroening, and David E. Bresler, "An Experimental Evaluation of Auricular Diagnosis: The Somatotopic Mapping of Musculoskeletal Pain at Ear Acupuncture Points," *Pain* 8 (1980), pp. 217–29.

73. Private communication with author, September 24, 1988.

74. Terrence D. Oleson and Richard J. Kroening, "Rapid Narcotic Detoxification in Chronic Pain Patients Treated with Auricular Electroacupuncture and Naloxone," *International Journal of the Addictions* 20, no. 9 (1985), pp. 1347–60.

75. Richard Leviton, "The Holographic Body," *East West* 18, no. 8 (August 1988), p. 42.

76. Ibid., p. 45.

77. Ibid., pp. 36–47.

78. "Fingerprints, a Clue to Senility," *Science Digest* 91, no. 11 (November 1983), p. 91.
79. Michael Meyer, "The Way the Whorls Turn," *Newsweek* (February 13, 1989), p. 73.

5. A POCKETFUL OF MIRACLES

1. D. Scott Rogo, *Miracles* (New York: Dial Press, 1982), p. 79.
2. Ibid., p. 58; see also, Herbert Thurston, *The Physical Phenomena of Mysticism* (London: Burns Oates, 1952); and A. P. Schimberg, *The Story of Therese Neumann* (Milwaukee, Wis.: Bruce Publishing Co., 1947).
3. David J. Bohm, "A New Theory of the Relationship of Mind and Matter," *Journal of the American Society for Psychical Research* 80, no. 2 (April 1986), p. 128.
4. Ibid., p. 132.
5. Robert G. Jahn and Brenda J. Dunne, *Margins of Reality: The Role of Consciousness in the Physical World* (New York: Harcourt Brace Jovanovich, 1987), pp. 91–123.
6. Ibid., p. 144.
7. Private communication with author, December 16, 1988.
8. Jahn and Dunne, *Margins*, p. 142.
9. Private communication with author, December 16, 1988.
10. Private communication with author, December 16, 1988.
11. Steve Fishman, "Questions for the Cosmos," *New York Times Magazine* (November 26, 1989), p. 55.
12. Private communication with author, November 25, 1988.
13. Rex Gardner, "Miracles of Healing in Anglo-Celtic Northumbria as Recorded by the Venerable Bede and His Contemporaries: A Reappraisal in the Light of Twentieth-Century Experience," *British Medical Journal* 287 (December 1983), p. 1931.
14. Max Freedom Long, *The Secret Science behind Miracles* (New York: Robert Collier Publications, 1948), pp. 191–92.
15. Louis-Basile Carre de Montgeron, *La Verité des Miracles* (Paris: 1737), vol. i, p. 380, as quoted in H. P. Blavatsky, *Isis Unveiled*, vol. i (New York: J. W. Bouton, 1877), p. 374.
16. Ibid., p. 374.
17. B. Robert Kreiser, *Miracles, Convulsions, and Ecclesiastical Politics in Early Eighteenth-Century Paris* (Princeton, N. J.: Princeton University Press, 1978), pp. 260–61.
18. Charles Mackey, *Extraordinary Popular Delusions and the Madness of Crowds* (London: 1841), p. 318.
19. Kreiser, *Miracles*, p. 174.

20. Stanislav Grof, *Beyond the Brain* (Albany, N.Y.: State University of New York Press, 1985), p. 91.

21. Long, *Secret Science*, pp. 31–39.

22. Frank Podmore, *Mediums of the Nineteenth Century*, vol. 2 (New Hyde Park, N.Y.: University Books, 1963), p. 264.

23. Vincent H. Gaddis, *Mysterious Fires and Lights* (New York: Dell, 1967), pp. 114–15.

24. Blavatsky, *Isis*, p. 370.

25. Podmore, *Mediums*, p. 264.

26. Will and Ariel Durant, *The Age of Louis XIV*, vol. XIII (New York: Simon & Schuster, 1963), p. 73.

27. Franz Werfel, *The Song of Bernadette* (Garden City, N.Y.: Sun Dial Press, 1944), pp. 326–27.

28. Gaddis, *Mysterious Fires*, pp. 106–7.

29. Ibid., p. 106.

30. Berthold Schwarz, "Ordeals by Serpents, Fire, and Strychnine," *Psychiatric Quarterly* 34 (1960), pp. 405–29.

31. Private communication with author, July 17, 1989.

32. Karl H. Pribram, "The Implicate Brain," in *Quantum Implications*, ed. Basil J. Hiley and F. David Peat (London: Routledge & Kegan Paul, 1987), p. 367.

33. Private communication with author, February 8, 1989; see also, Karl H. Pribram, "The Cognitive Revolution and Mind/Brain Issues," *American Psychologist* 41, no. 5 (May 1986), pp. 507–19.

34. Private communication with author, November 25, 1988.

35. Gordon G. Globus, "Three Holonomic Approaches to the Brain," in *Quantum Implications*, ed. Basil J. Hiley and F. David Peat (London: Routledge & Kegan Paul, 1987), pp. 372–85; see also, Judith Hooper and Dick Teresi, *The Three-Pound Universe* (New York: Dell, 1986), pp. 295–300.

36. Private communication with author, December 16, 1988.

37. Malcolm W. Browne, "Quantum Theory: Disturbing Questions Remain Unresolved," *New York Times* (February 11, 1986), p. C3.

38. Ibid.

39. Jahn and Dunne, *Margins*, pp. 319–20; see also, Dietrick E. Thomsen, "Anomalons Get More and More Anomalous," *Science News* 125 (February 25, 1984).

40. Christine Sutton, "The Secret Life of the Neutrino," *New Scientist* 117, no. 1595 (January 14, 1988), pp. 53–57; see also, "Soviet Neutrinos Have Mass," *New Scientist* 105, no. 1446 (March 7, 1985), p. 23; and Dietrick E. Thomsen, "Ups and Downs of Neutrino Oscillation," *Science News* 117, no. 24 (June 14, 1980), pp. 377–83.

41. S. Edmunds, *Hypnotism and the Supernormal* (London: Aquarian Press, 1967), as quoted in *Supernature*, Lyall Watson (New York: Bantam Books, 1973), p. 236.

42. Leonid L. Vasiliev, *Experiments in Distant Influence* (New York: E. P. Dutton, 1976).

43. See Russell Targ and Harold Puthoff, *Mind-Reach* (New York: Delacorte Press, 1977).

44. Fishman, *New York Times Magazine*, p. 55; see also, Jahn and Dunne, *Margins*, p. 187.

45. Charles Tart, "Physiological Correlates of Psi Cognition," *International Journal of Neuropsychiatry* 5, no. 4 (1962).

46. Targ and Puthoff, *Mind-Reach*, pp. 130–33.

47. E. Douglas Dean, "Plethysmograph Recordings of ESP Responses," *International Journal of Neuropsychiatry* 2 (September 1966).

48. Charles T. Tart, "Psychedelic Experiences Associated with a Novel Hypnotic Procedure, Mutual Hypnosis," in *Altered States of Consciousness*, Charles T. Tart (New York: John Wiley & Sons, 1969), pp. 291–308.

49. Ibid.

50. John P. Briggs and F. David Peat, *Looking Glass Universe* (New York: Simon & Schuster, 1984), p. 87.

51. Targ and Puthoff, *Mind-Reach*, pp. 130–33.

52. Russell Targ, et al., *Research in Parapsychology* (Metuchen, N.J.: Scarecrow, 1980).

53. Bohm, *Journal of the American Society for Psychical Research*, p. 132.

54. Jahn and Dunne, *Margins*, pp. 257–59.

55. Gardner, *British Medical Journal*, p. 1930.

56. Lyall Watson, *Beyond Supernature* (New York: Bantam Books, 1988), pp. 189–91.

57. A. R. G. Owen, *Can We Explain the Poltergeist* (New York: Garrett Publications, 1964).

58. Erlendur Haraldsson, *Modern Miracles: An Investigative Report on Psychic Phenomena Associated with Sathya Sai Baba* (New York: Fawcett Columbine Books, 1987), pp. 26–27.

59. Ibid., pp. 35–36.

60. Ibid., p. 290.

61. Paramahansa Yogananda, *Autobiography of a Yogi* (Los Angeles: Self-Realization Fellowship, 1973), p. 134.

62. Rogo, *Miracles*, p. 173.

63. Lyall Watson, *Gifts of Unknown Things* (New York: Simon & Schuster, 1976), pp. 203–4.

64. Private communication with author, February 9, 1989.
65. Private communication with author, October 17, 1988.
66. Private communication with author, December 16, 1988.
67. Judith Hooper and Dick Teresi, *The Three-Pound Universe* (New York: Dell, 1986), p. 300.
68. Carlos Castaneda, *Tales of Power* (New York: Simon & Schuster, 1974), p. 100.
69. Marilyn Ferguson, "Karl Pribram's Changing Reality," in *The Holographic Paradigm*, ed. Ken Wilber (Boulder, Colo.: New Science Library, 1982), p. 24.
70. Erlendur Haraldsson and Loftur R. Gissurarson, *The Icelandic Physical Medium: Indridi Indridason* (London: Society for Psychical Research, 1989).

6. SEEING HOLOGRAPHICALLY

1. Karl Pribram, "The Neurophysiology of Remembering," *Scientific American* 220 (January 1969), pp. 76–78.
2. Judith Hooper, "Interview: Karl Pribram," *Omni* 5, no. 1 (October 1982), p. 172.
3. Wil van Beek, *Hazrat Inayat Khan* (New York: Vantage Press, 1983), p. 135.
4. Barbara Ann Brennan, *Hands of Light* (New York: Bantam Books, 1987), pp. 3–4.
5. Ibid., p. 4.
6. Ibid., cover quote.
7. Ibid., cover quote.
8. Ibid., p. 26.
9. Private communication with author, November 13, 1988.
10. Shafica Karagulla, *Breakthrough to Creativity* (Marina Del Rey, Calif.: DeVorss, 1967), p. 61.
11. Ibid., pp. 78–79.
12. W. Brugh Joy, *Joy's Way* (Los Angeles: J. P. Tarcher, 1979), pp. 155–56.
13. Ibid., p. 48.
14. Michael Crichton, *Travels* (New York: Knopf, 1988), p. 262.
15. Ronald S. Miller, "Bridging the Gap: An Interview with Valerie Hunt," *Science of Mind* (October 1983), p. 12.
16. Private communication with author, February 7, 1990.
17. Ibid.
18. Ibid.
19. Ibid.

20. Valerie V. Hunt, "Infinite Mind," *Magical Blend,* no. 25 (January 1990), p. 22.
21. Private communication with author, October 28, 1988.
22. Robert Temple, "David Bohm," *New Scientist* (November 11, 1982), p. 362.
23. Private communication with author, November 13, 1988.
24. Private communication with author, October 18, 1988.
25. Private communication with author, November 13, 1988.
26. Ibid.
27. Ibid.
28. George F. Dole, *A View from Within* (New York: Swedenborg Foundation, 1985), p. 26.
29. George F. Dole, "An Image of God in a Mirror," in *Emanuel Swedenborg: A Continuing Vision,* ed. Robin Larsen (New York: Swedenborg Foundation, 1988), p. 376.
30. Brennan, *Hands,* p. 26.
31. Private communication with author, September 13, 1988.
32. Karagulla, *Breakthrough,* p. 39.
33. Ibid., p. 132.
34. D. Scott Rogo, "Shamanism, ESP, and the Paranormal," in *Shamanism,* ed. Shirley Nicholson (Wheaton, Ill.: Theosophical Publishing House, 1987), p. 135.
35. Michael Harner and Gary Doore, "The Ancient Wisdom in Shamanic Cultures," in *Shamanism,* ed. Shirley Nicholson (Wheaton, Ill.: Theosophical Publishing House, 1987), p. 10.
36. Michael Harner, *The Way of the Shaman* (New York: Harper & Row, 1980), p. 17.
37. Richard Gerber, *Vibrational Medicine* (Santa Fe, N.M.: Bear & Co., 1988), p. 115.
38. Ibid., p. 154.
39. William A. Tiller, "Consciousness, Radiation, and the Developing Sensory System," as quoted in *The Psychic Frontiers of Medicine,* ed. Bill Schul (New York: Ballantine Books, 1977), p. 95.
40. Ibid., p. 94.
41. Hiroshi Motoyama, *Theories of the Chakras* (Wheaton, Ill.: Theosophical Publishing House, 1981), p. 239.
42. Richard M. Restak, "Is Free Will a Fraud?" *Science Digest* (October 1983), p. 52.
43. Ibid.
44. Private communication with author, February 7, 1990.
45. Private communication with author, November 13, 1988.

7. TIME OUT OF MIND

1. See Stephan A. Schwartz, *The Secret Vaults of Time* (New York: Grosset & Dunlap, 1978); Stanislaw Poniatowski, "Parapsychological Probing of Prehistoric Cultures," in *Psychic Archaeology*, ed. J. Goodman (New York: G. P. Putnam & Sons, 1977); and Andrzey Borzmowski, "Experiments with Ossowiecki," *International Journal of Parapsychology* 7, no. 3 (1965), pp. 259–84.

2. J. Norman Emerson, "Intuitive Archaeology," *Midden* 5, no. 3 (1973).

3. J. Norman Emerson, "Intuitive Archaeology: A Psychic Approach," *New Horizon* 1, no. 3 (1974), p. 14.

4. Jack Harrison Pollack, *Croiset the Clairvoyant* (New York: Doubleday, 1964).

5. Lawrence LeShan, *The Medium, the Mystic, and the Physicist* (New York: Ballantine Books, 1974), pp. 30–31.

6. Stephan A. Schwartz, *The Secret Vaults of Time* (New York: Grosset & Dunlap, 1978), pp. 226–37; see also Clarence W. Weiant, "Parapsychology and Anthropology," *Manas* 13, no. 15 (1960).

7. Schwartz, op. cit., pp. x and 314.

8. Private communication with author, October 28, 1988.

9. Private communication with author, October 18, 1988.

10. See Glenn D. Kittler, *Edgar Cayce on the Dead Sea Scrolls* (New York: Warner Books, 1970).

11. Marilyn Ferguson, "Quantum Brain-Action Approach Complements Holographic Model," *Brain-Mind Bulletin*, updated special issue (1978), p. 3.

12. Edmund Gurney, F. W. H. Myers, and Frank Podmore, *Phantasms of the Living* (London: Trubner's, 1886).

13. See J. Palmer, "A Community Mail Survey of Psychic Experiences," *Journal of the American Society for Psychical Research* 73 (1979), pp. 221–51; H. Sidgwick and committee, "Report on the Census of Hallucinations," *Proceedings of the Society for Psychical Research* 10 (1894), pp. 25–422; and D. J. West, "A Mass-Observation Questionnaire on Hallucinations," *Journal of the Society for Psychical Research* 34 (1948), pp. 187–96.

14. W. Y. Evans-Wentz, *The Fairy-Faith in Celtic Countries* (Oxford: Oxford University Press, 1911), p. 485.

15. Ibid., p. 123.

16. Charles Fort, *New Lands* (New York: Boni & Liveright, 1923), p. 111.

17. See Max Freedom Long, *The Secret Science behind Miracles* (Tarrytown, N.Y.: Robert Collier Publications, 1948), pp. 206–8.

18. Editors of Time-Life Books, *Ghosts* (Alexandria, Va.: Time-Life Books, 1984), p. 75.

19. Editors of Reader's Digest, *Strange Stories, Amazing Facts* (Pleasant-ville, N.Y.: Reader's Digest Association, 1976), pp. 384–85.

20. J. B. Rhine, "Experiments Bearing on the Precognition Hypothesis: III. Mechanically Selected Cards," *Journal of Parapsychology* 5 (1941).

21. Helmut Schmidt, "Psychokinesis," in *Psychic Exploration: A Challenge to Science,* ed. Edgar Mitchell and John White (New York: G. P. Putnam's Sons, 1974), pp. 179–93.

22. Montague Ullman, Stanley Krippner, and Alan Vaughan, *Dream Telepathy* (New York: Macmillan, 1973).

23. Russell Targ and Harold Puthoff, *Mind-Reach* (New York: Delacorte Press, 1977), p. 116.

24. Robert G. Jahn and Brenda J. Dunne, *Margins of Reality* (New York: Harcourt Brace Jovanovich, 1987), pp. 160, 185.

25. Jule Eisenbud, "A Transatlantic Experiment in Precognition with Gerard Croiset," *Journal of American Society of Psychological Research* 67 (1973), pp. 1–25; see also W. H. C. Tenhaeff, "Seat Experiments with Gerard Croiset," *Proceedings Parapsychology* 1 (1960), pp. 53–65; and U. Timm, "Neue Experiments mit dem Sensitiven Gerard Croiset," *Z. F. Parapsychologia und Grezgeb. dem Psychologia* 9 (1966), pp. 30–59.

26. Marilyn Ferguson, *Bulletin,* p. 4.

27. Personal communication with author, September 26, 1989.

28. David Loye, *The Sphinx and the Rainbow* (Boulder, Col.: Shambhala, 1983).

29. Bernard Gittelson, *Intangible Evidence* (New York: Simon & Schuster, 1987), p. 174.

30. Eileen Garrett, *My Life as a Search for the Meaning of Mediumship* (London: Ryder & Company, 1949), p. 179.

31. Edith Lyttelton, *Some Cases of Prediction* (London: Bell, 1937).

32. Louisa E. Rhine, "Frequency of Types of Experience in Spontaneous Precognition," *Journal of Parapsychology* 18, no. 2 (1954); see also "Precognition and Intervention," *Journal of Parapsychology* 19 (1955); and *Hidden Channels of the Mind* (New York: Sloane Associates, 1961).

33. E. Douglas Dean, "Precognition and Retrocognition," in *Psychic Exploration,* ed. Edgar D. Mitchell and John White (New York: G. P. Putnam's Sons, 1974), p. 163.

34. See A. Foster, "ESP Tests with American Indian Children," *Journal of Parapsychology* 7, no. 94 (1943); Dorothy H. Pope, "ESP Tests with Primitive People," *Parapsychology Bulletin* 30, no. 1 (1953); Ronald Rose and Lyndon Rose, "Psi Experiments with Australian Aborigines," *Journal of Parapsychology* 15, no. 122 (1951); Robert L. Van de Castle, "Anthropology and Psychic Research," in *Psychic Exploration,* ed. Edgar D. Mitchell and John White (New York: G. P. Putnam's Sons, 1974); and Robert L. Van de Castle, "Psi Abilities in Primitive Groups," *Proceedings of the Parapsychological Association* 7, no. 97 (1970).

35. Ian Stevenson, "Precognition of Disasters," *Journal of the American Society for Psychical Research* 64, no. 2 (1970).

36. Karlis Osis and J. Fahler, "Space and Time Variables in ESP," *Journal of the American Society for Psychical Research* 58 (1964).

37. Alexander P. Dubrov and Veniamin N. Pushkin, *Parapsychology and Contemporary Science*, trans. Aleksandr Petrovich (New York: Consultants Bureau, 1982), pp. 93–104.

38. Arthur Osborn, *The Future Is Now: The Significance of Precognition* (New York: University Books, 1961).

39. Ian Stevenson, "A Review and Analysis of Paranormal Experiences Connected with the Sinking of the *Titanic*," *Journal of the American Society for Psychical Research* 54 (1960), pp. 153–71; see also Ian Stevenson, "Seven More Paranormal Experiences Associated with the Sinking of the *Titanic*," *Journal of the American Society for Psychical Research* 59 (1965), pp. 211–25.

40. Loye, *Sphinx*, pp. 158–65.

41. Private communication with author, October 28, 1988.

42. Gittelson, *Evidence*, p. 175.

43. Ibid., p. 125.

44. Long, op. cit., p. 165.

45. Shafica Karagulla, *Breakthrough to Creativity* (Marina Del Rey, Calif.: DeVorss, 1967), p. 206.

46. According to H. N. Banerjee, in *Americans Who Have Been Reincarnated* (New York: Macmillan Publishing Company, 1980), p. 195, one study done by James Parejko, a professor of philosophy at Chicago State University, revealed that 93 out of 100 hypnotized volunteers produced knowledge of a possible previous existence; Whitton himself has found that *all* of his hypnotizable subjects were able to recall such memories.

47. M. Gerald Edelstein, *Trauma, Trance and Transformation* (New York: Brunner/Mazel, 1981).

48. Michael Talbot, "Lives between Lives: An Interview with Dr. Joel Whitton" *Omni WholeMind Newsletter* 1, no. 6 (May 1988), p. 4.

49. Joel L. Whitton and Joe Fisher, *Life between Life* (New York: Doubleday, 1986), pp. 116–27.

50. Ibid., p. 154.

51. Ibid., p. 156.

52. Private communication with author, November 9, 1987.

53. Whitton and Fisher, *Life*, p. 43.

54. Ibid., p. 47.

55. Ibid., pp. 152–53.

56. Ibid., p. 52.

57. William E. Cox, "Precognition: An Analysis I and II," *Journal of the American Society for Psychical Research* 50 (1956).

58. Whitton and Fisher, *Life*, p. 186.

59. See Ian Stevenson, *Twenty Cases Suggestive of Reincarnation* (Charlottesville, Va.: University Press of Virginia, 1974); *Cases of the Reincarnation Type* (Charlottesville, Va.: University Press of Virginia, 1974), vols. 1–4; and *Children Who Remember Their Past Lives* (Charlottesville, Va.: University Press of Virginia, 1987).

60. See references above.

61. Ian Stevenson, *Children Who Remember Previous Lives* (Charlottesville, Va.: University Press of Virginia, 1987), pp. 240–43.

62. Ibid., pp. 259–60.

63. Stevenson, *Twenty Cases*, p. 180.

64. Ibid., pp. 196, 233.

65. Ibid., p. 92.

66. Sylvia Cranston and Carey Williams, *Reincarnation: A New Horizon in Science, Religion, and Society* (New York: Julian Press, 1984), p. 67.

67. Ibid., p. 260.

68. Ian Stevenson, "Some Questions Related to Cases of the Reincarnation Type," *Journal of the American Society for Psychical Research* (October 1974), p. 407.

69. Stevenson, *Children*, p. 255.

70. *Journal of the American Medical Association* (December 1, 1975), as quoted in Cranston and Williams, *Reincarnation*, p. x.

71. J. Warneck, *Die Religion der Batak* (Gottingen, 1909), as quoted in Holger Kalweit, *Dreamtime and Inner Space: The World of the Shaman* (Boulder, Colo.: Shambhala, 1984), p. 23.

72. Basil Johnston, *Und Manitu erschuf die Welt. Mythen and Visionen der Ojibwa* (Cologne: 1979), as quoted in Holger Kalweit, *Dreamtime and Inner Space: The World of the Shaman* (Boulder, Colo.: Shambhala, 1984), p. 25.

73. Long, op. cit., pp. 165–69.

74. Ibid., p. 193.

75. John Blofeld, *The Tantric Mysticism of Tibet* (New York: E. P. Dutton, 1970), p. 84; see also Alexandra David-Neel, *Magic and Mystery in Tibet* (Baltimore, Md.: Penguin Books, 1971), p. 293.

76. Henry Corbin, *Creative Imagination in the Sufism of Ibn 'Arabi*, trans. Ralph Manheim (Princeton, N.J.: Princeton University Press, 1969), pp. 221–36.

77. Hugh Lynn Cayce, *The Edgar Cayce Reader. Vol. II* (New York: Paperback Library, 1969), pp. 25–26; see also Noel Langley, *Edgar Cayce on Reincarnation* (New York: Warner Books, 1967), p. 43.

78. Paramahansa Yogananda, *Man's Eternal Quest* (Los Angeles: Self-Realization Fellowship, 1982), p. 238.

79. Thomas Byron, *The Dhammapada: The Sayings of Buddha* (New York: Vintage Books, 1976), p. 13.

80. Swami Prabhavananda and Frederick Manchester, trans., The Upanishads (Hollywood, Calif.: Vedanta Press, 1975), p. 177.

81. Iamblichus, *The Egyptian Mysteries*, trans. Alexander Wilder (New York: Metaphysical Publications, 1911), pp. 122, 175, 259–60.

82. Matthew 7: 7, 17, 20.

83. Rabbi Adin Steinsaltz, *The Thirteen-Petaled Rose* (New York: Basic Books, 1980), pp. 64–65.

84. Jean Houston, *The Possible Human* (Los Angeles: J. P. Tarcher, 1982), pp. 200–5.

85. Mary Orser and Richard A. Zarro, *Changing Your Destiny* (San Francisco: Harper & Row, 1989), p. 213.

86. Florence Graves, "The Ultimate Frontier: Edgar Mitchell, the Astronaut-Turned-Philosopher Explores Star Wars, Spirituality, and How We Create Our Own Reality," *New Age* (May/June 1988), p. 87.

87. Helen Wambach, *Reliving Past Lives* (New York: Harper & Row, 1978), p. 116.

88. Ibid., pp. 128–34.

89. Chet B. Snow and Helen Wambach, *Mass Dreams of the Future* (New York: McGraw-Hill, 1989), p. 218.

90. Henry Reed, "Reaching into the Past with Mind over Matter," *Venture Inward* 5, no. 3 (May/June 1989), p. 6.

91. Anne Moberly and Eleanor Jourdain, *An Adventure* (London: Faber, 1904).

92. Andrew Mackenzie, *The Unexplained* (London: Barker, 1966), as quoted in Ted Holiday, *The Goblin Universe* (St. Paul, Minn.: Llewellyn Publications, 1986), p. 96.

93. Gardner Murphy and H. L. Klemme, "Unfinished Business," *Journal of the American Society for Psychical Research* 60, no. 4 (1966), p. 5.

8. TRAVELING IN THE SUPERHOLOGRAM

1. Dean Shields, "A Cross-Cultural Study of Beliefs in out-of-the-Body Experiences," *Journal of the Society for Psychical Research* 49 (1978), pp. 697–741.

2. Erika Bourguignon, "Dreams and Altered States of Consciousness in Anthropological Research," in *Psychological Anthropology*, ed. F. L. K. Hsu (Cambridge, Mass.: Schenkman, 1972), p. 418.

3. Celia Green, *Out-of-the-Body Experiences* (Oxford, England: Institute of Psychophysical Research, 1968).

4. D. Scott Rogo, *Leaving the Body* (New York: Prentice-Hall, 1983), p. 5.
5. Ibid.
6. Stuart W. Twemlow, Glen O. Gabbard, and Fowler C. Jones, "The Out-of-Body Experience: I, Phenomenology; II, Psychological Profile; III, Differential Diagnosis" (Papers delivered at the 1980 Convention of the American Psychiatric Association). See also Twemlow, Gabbard, and Jones, "The Out-of-Body Experience: A Phenomenological Typology Based on Questionnaire Responses," *American Journal of Psychiatry* 139 (1982), pp. 450–55.
7. Ibid.
8. Bruce Greyson and C. P. Flynn, *The Near-Death Experience* (Chicago: Charles C. Thomas, 1984), as quoted in Stanislov Grof, *The Adventure of Self-Discovery* (Albany, N.Y.: SUNY Press, 1988), pp. 71–72.
9. Michael B. Sabom, *Recollections of Death* (New York: Harper & Row, 1982), p. 184.
10. Jean-Noel Bassior, "Astral Travel," *New Age Journal* (November/December 1988), p. 46.
11. Charles Tart, "A Psychophysiological Study of Out-of-the-Body Experiences in a Selected Subject," *Journal of the American Society for Psychical Research* 62 (1968), pp. 3–27.
12. Karlis Osis, "New ASPR Research on Out-of-the-Body Experiences," *Newsletter of the American Society for Psychical Research* 14 (1972); see also Karlis Osis, "Out-of-Body Research at the American Society for Psychical Research," in *Mind beyond the Body*, ed. D. Scott Rogo (New York: Penguin, 1978), pp. 162–69.
13. D. Scott Rogo, *Psychic Breakthroughs Today* (Wellingborough, Great Britain: Aquarian Press, 1987), pp. 163–64.
14. J. H. M. Whiteman, *The Mystical Life* (London: Faber & Faber, 1961).
15. Robert A. Monroe, *Journeys Out of the Body* (New York: Anchor Press/Doubleday, 1971), p. 183.
16. Robert A. Monroe, *Far Journeys* (New York: Doubleday, 1985), p. 64.
17. David Eisenberg, with Thomas Lee Wright, *Encounters with Qi* (New York: Penguin, 1987), pp. 79–87.
18. Frank Edwards, "People Who Saw without Eyes," *Strange People* (London: Pan Books, 1970).
19. A. Ivanov, "Soviet Experiments in Eyeless Vision," *International Journal of Parapsychology* 6 (1964); see also M. M. Bongard and M. S. Smirnov, "About the 'Dermal Vision' of R. Kuleshova," *Biophysics* 1 (1965).
20. A. Rosenfeld, "Seeing Colors with the Fingers," *Life* (June 12, 1964); for a more extensive report of Kuleshova and "eyeless sight" in general, see Sheila Ostrander and Lynn Schroeder, *Psychic Discoveries Behind the Iron Curtain* (New York: Bantam Books, 1970), pp. 170–85.
21. Rogo, *Psychic Breakthroughs*, p. 161.

22. Ibid.

23. Janet Lee Mitchell, *Out-of-Body Experiences* (New York: Ballantine Books, 1987), p. 81.

24. August Strindberg, *Legends* (1912 edition), as quoted in Colin Wilson, *The Occult* (New York: Vintage Books, 1973), pp. 56–57.

25. Monroe, *Journeys Out of the Body*, p. 184.

26. Whiteman, *Mystical Life*, as quoted in Mitchell, *Experiences*, p. 44.

27. Karlis Osis and Erlendur Haraldsson, "Deathbed Observations by Physicians and Nurses: A Cross-Cultural Survey," *The Journal of the American Society for Psychical Research* 71 (July 1977), pp. 237–59.

28. Raymond A. Moody, Jr., with Paul Perry, *The Light Beyond* (New York: Bantam Books, 1988), pp. 14–15.

29. Ibid.

30. Elisabeth Kubler-Ross, *On Children and Death* (New York: Macmillan, 1983), p. 208.

31. Kenneth Ring, *Life at Death* (New York: Quill, 1980), pp. 238–39.

32. Kubler-Ross, *Children*, p. 210.

33. Moody and Perry, *Light*, pp. 103–7.

34. Ibid., p. 151.

35. George Gallup, Jr., with William Proctor, *Adventures in Immortality* (New York: McGraw-Hill, 1982), p. 31.

36. Ring, *Life at Death*, p. 98.

37. Ibid., pp. 97–98.

38. Ibid., p. 247.

39. Private communication with author, May 24, 1990.

40. F. W. H. Myers, *Human Personality and Its Survival of Bodily Death* (London: Longmans, Green & Co., 1904), pp. 315–21.

41. Ibid.

42. Moody and Perry, *Light*, p. 8.

43. Joel L. Whitton and Joe Fisher, *Life between Life* (New York: Doubleday, 1986), p. 32.

44. Michael Talbot, "Lives between Lives: An Interview with Joel Whitton," *Omni WholeMind Newsletter* 1, no. 6 (May 1988), p. 4.

45. Private communication with author, November 9, 1987.

46. Whitton and Fisher, *Life between Life*, p. 35.

47. Myra Ka Lange, "To the Top of the Universe," *Venture Inward* 4, no. 3 (May/June 1988), p. 42.

48. F. W. H. Myers, *Human Personality*.

49. Moody and Perry, *Light*, p. 129.

50. Raymond A. Moody, Jr., *Reflections on Life after Life* (New York: Bantam Books, 1978), p. 38.

51. Whitton and Fisher, *Life between Life*, p. 39.

52. Raymond A. Moody, Jr., *Life after Life* (New York: Bantam Books, 1976), p. 68.

53. Moody, *Reflections on Life after Life*, p. 35.

54. The 1821 NDEer was the mother of the English writer Thomas De Quincey and the incident is described in his *Confessions of an English Opium Eater with Its Sequels Suspiria De Profundis and The English Mail-Coach*, ed. Malcolm Elwin (London: Macdonald & Co., 1956), pp. 511–12.

55. Whitton and Fisher, *Life between Life*, pp. 42–43.

56. Moody and Perry, *Light*, p. 50.

57. Ibid., p. 35.

58. Kenneth Ring, *Heading toward Omega* (New York: William Morrow, 1985), pp. 58–59.

59. See Ring, *Heading toward Omega*, p. 199; Moody, *Reflections on Life after Life*, pp. 9–14; and Moody and Perry, *Light*, p. 35.

60. Moody and Perry, *Light*, p. 35.

61. Monroe, *Far Journeys*, p. 73.

62. Ring, *Life at Death*, p. 248.

63. Ibid., p. 242.

64. Moody, *Life after Life*, p. 75.

65. Moody and Perry, *Light*, p. 13.

66. Ring, *Heading toward Omega*, pp. 186–87.

67. Moody and Perry, *Light*, p. 22.

68. Ring, *Heading toward Omega*, pp. 217–18.

69. Moody and Perry, *Light*, p. 34.

70. Ian Stevenson, *Children Who Remember Previous Lives* (Charlottesville, Va.: University Press of Virginia, 1987), p. 110.

71. Whitton and Fisher, *Life between Life*, p. 43.

72. Wil van Beek, *Hazrat Inayat Khan* (New York: Vantage Press, 1983), p. 29.

73. Monroe, *Journeys Out of the Body*, pp. 101–15.

74. See Leon S. Rhodes, "Swedenborg and the Near-Death Experience," in *Emanuel Swedenborg: A Continuing Vision*, ed. Robin Larsen et al. (New York: Swedenborg Foundation, 1988), pp. 237–40.

75. Wilson Van Dusen, *The Presence of Other Worlds* (New York: Swedenborg Foundation, 1974), p. 75.

76. Emanuel Swedenborg, *The Universal Human and Soul-Body Interaction*, ed. and trans. George F. Dole (New York: Paulist Press, 1984), p. 43.

77. Ibid.

78. Ibid., p. 156.

79. Ibid., p. 45.

80. Ibid., p. 161.

81. George F. Dole, "An Image of God in a Mirror," in *Emanuel Sweden-borg: A Continuing Vision*, ed. Robin Larsen et al. (New York: Swedenborg Foundation, 1988), pp. 374–81.

82. Ibid.

83. Theophilus Parsons, *Essays* (Boston: Otis Clapp, 1845), p. 225.

84. Henry Corbin, *Mundus Imaginalis* (Ipswich, England: Golgonooza Press, 1976), p. 4.

85. Ibid., p. 7.

86. Ibid., p. 5.

87. Kubler-Ross, *Children*, p. 222.

88. Private communication with author, October 28, 1988.

89. Paramahansa Yogananda, *Autobiography of a Yogi* (Los Angeles: Self-Realization Fellowship, 1973), p. viii.

90. Ibid., pp. 475–97.

91. Satprem, *Sri Aurobindo or the Adventure of Consciousness* (New York: Institute for Evolutionary Research, 1984), p. 195.

92. Ibid., p. 219.

93. E. Nandisvara Nayake Thero, "The Dreamtime, Mysticism, and Liberation: Shamanism in Australia," in *Shamanism*, ed. Shirley Nicholson (Wheaton, Ill.: Theosophical Publishing House, 1987), pp. 223–32.

94. Holger Kalweit, *Dreamtime and Inner Space* (Boston: Shambhala Publications, 1984), pp. 12–13.

95. Michael Harner, *The Way of the Shaman* (New York: Harper & Row, 1980), pp. 1–8.

96. Kalweit, *Dreamtime*, pp. 13, 57.

97. Ring, *Heading toward Omega*, pp. 143–64.

98. Ibid., pp. 114–20.

99. Bruce Greyson, "Increase in Psychic and Psi-Related Phenomena Following Near-Death Experiences," *Theta*, as quoted in Ring, *Heading toward Omega*, p. 180.

100. Jeff Zaleski, "Life after Death: Not Always Happily-Ever-After," *Omni WholeMind Newsletter* 1, no. 10 (September 1988), p. 5.

101. Ring, *Heading toward Omega*, p. 50.

102. John Gliedman, "Interview with Brian Josephson," *Omni* 4, no. 10 (July 1982), pp. 114–16.

103. P. C. W. Davies, "The Mind-Body Problem and Quantum Theory," in *Proceedings of the Symposium on Consciousness and Survival*, ed. John S. Spong (Sausalito, Calif.: Institute of Noetic Sciences, 1987), pp. 113–14.

104. Candace Pert, *Neuropeptides, the Emotions and Bodymind in Proceedings of the Symposium on Consciousness and Survival,* ed. John S. Spong (Sausalito, Calif.: Institute of Noetic Sciences, 1987), pp. 113–14.

105. David Bohm and Renee Weber, "Nature as Creativity," *ReVision* 5, no. 2 (Fall 1982), p. 40.

106. Private communication with author, November 9, 1987.

107. Monroe, *Journeys Out of the Body,* pp. 51 and 70.

108. Dole, in *Emanuel Swedenborg,* p. 44.

109. Whitton and Fisher, *Life between Life,* p. 45.

110. See, for example, Moody, *Reflections on Life after Life,* pp. 13–14; and Ring, *Heading toward Omega,* pp. 71–72.

111. Edwin Bernbaum, *The Way to Shambhala* (New York: Anchor Books, 1980), pp. xiv, 3–5.

112. Moody, *Reflections on Life after Life,* p. 14; and Ring, *Heading toward Omega,* p. 71.

113. W. Y. Evans-Wentz, *The Fairy-Faith in Celtic Countries* (Oxford: Oxford University Press, 1911), p. 61.

114. Monroe, *Journeys Out of the Body,* pp. 50–51.

115. Jacques Vallee, *Passport to Magonia* (Chicago: Henry Regnery Co., 1969), p. 134.

116. Private communication with author, November 3, 1988.

117. D. Scott Rogo, *Miracles* (New York: Dial Press, 1982), pp. 256–57.

118. Michael Talbot, "UFOs: Beyond Real and Unreal," in *Gods of Aquarius,* ed. Brad Steiger (New York: Harcourt Brace Jovanovich, 1976), pp. 28–33.

119. Jacques Vallee, *Dimensions: A Casebook of Alien Contact* (Chicago: Contemporary Books, 1988), p. 259.

120. John G. Fuller, *The Interrupted Journey* (New York: Dial Press, 1966), p. 91.

121. Jacques Vallee, *Passport to Magonia,* pp. 160–62.

122. Talbot, in *Gods of Aquarius,* pp. 28–33.

123. Kenneth Ring, "Toward an Imaginal Interpretation of 'UFO Abductions,'" *ReVision* 11, no. 4 (Spring 1989), pp. 17–24.

124. Personal communication with author, September 19, 1988.

125. Peter M. Rojcewicz, "The Folklore of the 'Men in Black': A Challenge to the Prevailing Paradigm," *ReVision* 11, no. 4 (Spring 1989), pp. 5–15.

126. Whitley Strieber, *Communion* (New York: Beech Tree Books, 1987), p. 295.

127. Carl Raschke, "UFOs: Ultraterrestrial Agents of Cultural Deconstruction," in *Cyberbiological Studies of the Imaginal Component in the UFO Contact Experience,* ed. Dennis Stillings (St. Paul, Minn.: Archaeus Project, 1989), p. 24.

128. Michael Grosso, "UFOs and the Myth of the New Age," in *Cyberbiological Studies of the Imaginal Component in the UFO Contact Experience*, ed. Dennis Stillings (St. Paul, Minn.: Archaeus Project, 1989), p. 81.

129. Raschke, in *Cyberbiological Studies*, p. 24.

130. Jacques Vallee, *Dimensions: A Casebook of Alien Contact* (Chicago: Contemporary Books, 1988), pp. 284–89.

131. John A. Wheeler, with Charles Misner and Kip S. Thorne, *Gravitation* (San Francisco: Freeman, 1973).

132. Strieber, *Communion*, p. 295.

133. Private communication with author, June 8, 1988.

9. RETURN TO THE DREAMTIME

1. John Blofeld, *The Tantric Mysticism of Tibet* (New York: E. P. Dutton, 1970), pp. 61–62.

2. Garma C. C. Chuang, *Teachings of Tibetan Yoga* (Secaucus, N.J.: Citadel Press, 1974), p. 26.

3. Blofeld, *Tantric Mysticism*, pp. 61–62.

4. Lobsang P. Lhalungpa, trans., *The Life of Milarepa* (Boulder, Colo.: Shambhala Publications, 1977), pp. 181–82.

5. Reginald Horace Blyth, *Games Zen Masters Play*, ed. Robert Sohl and Audrey Carr (New York: New American Library, 1976), p. 15.

6. Margaret Stutley, *Hinduism* (Wellingborough, England: Aquarian Press, 1985), pp. 9, 163.

7. Swami Prabhavananda and Frederick Manchester, trans., The Upanishads (Hollywood, Calif.: Vedanta Press, 1975), p. 197.

8. Sir John Woodroffe, *The Serpent Power* (New York: Dover, 1974), p. 33.

9. Stutley, *Hinduism*, p. 27.

10. Ibid., pp. 27–28.

11. Woodroffe, *Serpent Power*, pp. 29, 33.

12. Leo Schaya, *The Universal Meaning of the Kabbalah* (Baltimore, Md.: Penguin, 1973), p. 67.

13. Ibid.

14. Serge King, "The Way of the Adventurer," in *Shamanism*, ed. Shirley Nicholson (Wheaton, Ill.: Theosophical Publishing House, 1987), p. 193.

15. E. Nandisvara Nayake Thero, "The Dreamtime, Mysticism, and Liberation: Shamanism in Australia," in *Shamanism*, ed. Shirley Nicholson (Wheaton, Ill.: Theosophical Publishing House, 1987), p. 226.

16. Marcel Griaule, *Conversations with Ogotemmeli* (London: Oxford University Press, 1965), p. 108.

17. Douglas Sharon, *Wizard of the Four Winds: A Shaman's Story* (New York: Free Press, 1978), p. 49.

18. Henry Corbin, *Creative Imagination in the Sufism of Ibn 'Arabi*, trans. Ralph Manheim (Princeton, N.J.: Princeton University Press, 1969), p. 259.
19. Brian Brown, *The Wisdom of the Egyptians* (New York: Brentano's, 1923), p. 156.
20. Woodroffe, *Serpent Power*, p. 22.
21. John G. Neihardt, *Black Elk Speaks* (New York: Pocket Books, 1972), p. 36.
22. Tryon Edwards, *A Dictionary of Thought* (Detroit: F. B. Dickerson Co., 1901), p. 196.
23. Sir Charles Eliot, *Japanese Buddhism* (New York: Barnes & Noble, 1969), pp. 109–10.
24. Alan Watts, *Tao: The Watercourse Way* (New York: Pantheon Books, 1975), p. 35.
25. F. Franck, *Book of Angelus Silesius* (New York: Random House, 1976), as quoted in Stanislav Grof, *Beyond the Brain* (Albany, N.Y.: SUNY Press, 1985), p. 76.
26. "'Holophonic' Sound Broadcasts Directly to Brain," *Brain/Mind Bulletin* 8, no. 10 (May 30, 1983), p. 1.
27. "European Media See Holophony as Breakthrough," *Brain/Mind Bulletin* 8, no. 10 (May 30, 1983), p. 3.
28. Ilya Prigogine and Yves Elskens, "Irreversibility, Stochasticity and Non-Locality in Classical Dynamics," in *Quantum Implications*, ed. Basil J. Hiley and F. David Peat (London: Routledge & Kegan Paul, 1987), p. 214; see also "A Holographic Fit?" *Brain/Mind Bulletin* 4, no. 13 (May 21, 1979), p. 3.
29. Marcus S. Cohen, "Design of a New Medium for Volume Holographic Information Processing," *Applied Optics* 25, no. 14 (July 15, 1986), pp. 2288–94.
30. Dana Z. Anderson, "Coherent Optical Eigenstate Memory," *Optics Letters* 11, no. 1 (January 1986), pp. 56–58.
31. Willis W. Harman, "The Persistent Puzzle: The Need for a Basic Restructuring of Science," *Noetic Sciences Review*, no. 8 (Autumn 1988), p. 23.
32. "Interview: Brian L. Weiss, M.D.," *Venture Inward* 6, no. 4 (July/August 1990), pp. 17–18.
33. Private communication with author, November 9, 1987.
34. Stanley R. Dean, C. O. Plyler, Jr., and Michael L. Dean, "Should Psychic Studies Be Included in Psychiatric Education? An Opinion Survey," *American Journal of Psychiatry* 137, no. 10 (October 1980), pp. 1247–49.
35. Ian Stevenson, *Children Who Remember Previous Lives* (Charlottesville, Va.: University Press of Virginia, 1987), p. 9.

36. Alexander P. Dubrov and Veniamin N. Pushkin, *Parapsychology and Contemporary Science* (New York: Consultants Bureau, 1982), p. 13.

37. Harman, *Noetic Sciences Review*, p. 25.

38. Kenneth Ring, "Near-Death and UFO Encounters as Shamanic Initiations: Some Conceptual and Evolutionary Implications," *ReVision* 11, no. 3 (Winter 1989), p. 16.

39. Richard Daab and Michael Peter Langevin, "An Interview with Whitley Strieber," *Magical Blend* 25 (January 1990), p. 41.

40. Lytle Robinson, *Edgar Cayce's Story of the Origin and Destiny of Man* (New York: Berkley Medallion, 1972), pp. 34, 42.

41. From the Lankavatara Sutra as quoted by Ken Wilbur, "Physics, Mysticism, and the New Holographic Paradigm," in Ken Wilbur, *The Holographic Paradigm* (Boulder, Colo.: New Science Library, 1982), p. 161.

42. David Loye, *The Sphinx and the Rainbow* (Boulder, Colo.: Shambhala Publications, 1983), p. 156.

43. Terence McKenna, "New Maps of Hyperspace," *Magical Blend* 22 (April 1989), pp. 58, 60.

44. Daab and Langevin, *Magical Blend*, p. 41.

45. McKenna, *Magical Blend*, p. 60.

46. Emanuel Swedenborg, *The Universal Human and Soul-Body Interaction*, ed. and trans. George F. Dole (New York: Paulist Press, 1984), p. 54.

47. Joel L. Whitton and Joe Fisher, *Life between Life* (New York: Doubleday, 1986), pp. 45–46.

Index

of, 300; participation in, 191; of
subatomic particles, 140, 146, 284
Creative Visualization, Gawain, 222
Crichton, Michael, 174
Croiset, Gerard, 199–200, 207
Crookall, Robert, 230
Crown chakra, 166
Cultural beliefs, 101–2
Cultural differences in near-death
experiences, 256

Dajo, Mirin, 103–4
Dale, Ralph Alan, 115–16
Dalibard, Jean, 52–53
Davies, John, 204
Davies, Paul, 53, 79, 270
Death, views of, 2, 261
Descartes, Réne, 247
d'Espagnat, Bernard, 54
DeValois, Karen and Russell, 27–28
Disease, energy fields and, 188–89
Disembodied states, 246–48. *See also*
Out-of-body experiences
Disorder, 44–46
Dissipative structures, 293
Distribution: of information, 48; of
memory, 13–14, 30; of vision, 20
Divine intelligence, 285
Doctors: and auras, 171–74; and drug
effectiveness, 92
Dogon people, Sudan, 289
Dole, George F., 259
Don Juan (Yaqui shaman), 138, 155–56,
160
Dorsett, Sybil, 99
Dosage, and placebo effects, 91–92
Dossey, Larry, 30, 89–90, 197
Dreams, 60–65, 182; information in, 252;
location in, 234; lucid, 3, 65–66; and
objective reality, 80; out-of-body
experiences and, 272–73n;
precognitive, 206, 209–10; reality as,
285
Dreams of a Spirit-Seer, Kant, 257
Dreamtime and Inner Space, Kalweit,
195
Drugs, effectiveness of, 94–97, 99
Dryer, Carol, 169, 180–83, 186, 192, 284
Dubrov, Alexander P., 110, 297–98
Dunne, Brenda J., 5, 123–26, 139–40,
146, 207
Dychtwald, Ken, 57

Ear acupuncture, 113–15
Egyptian Book of the Dead, 240, 241
Eidetic memory, 23–24
Einstein, Albert, 35–39, 48
Eisenberg, David, 236

Eisenbud, Jule, 207
Electromyograms (EMGs), 174–76
Electrons, 33–34, 140, 159; Bohm's
ideas, 47, 48, 50, 122; in plasma, 38
Emerson, Norman, 199
Emotions, holographic record, 203
Empedocles, 290
Energy, in space, Bohm's ideas, 51–52
Energy fields, human, 165–93
Enfolded order. *See* Implicate order
Engrams, 11–13
Epileptics, brain studies, 12
EPR (Einstein-Podolsky-Rosen) paradox,
37
ESP. *See* Extrasensory perception
Estebany, Oscar, 172
Etheric body, 166, 170, 188–89
Evans-Wentz, W. Y., 203–4, 262
Evolution, psychic, 299–302
Experience, holographic idea, 84
Experiments in Distant Influence,
Vasiliev, 142
Explicate order, 46–48
External realities, 24–25
Extrasensory perception (ESP), 141–44,
210; dream experiments, 6, 61–62
Eye, blind spot, 163
Eyeless sight, 236–37

Fahler, J., 210
Fairies, 203–4
The Fairy-Faith in Celtic Countries,
Evans-Wentz, 204
Faith, beliefs and, 107–10
Familiar things, recognition of, 22–23
Far Journeys, Monroe, 233
Faster-than-light communication, 36–37,
53
Fasting state, 153–54, 256
Fa-Tsang, 291
Feinberg, Leonard, 136
Feinstein, Bertram, 191–92
Fenske, Elizabeth W., 245–46
The Final Choice, Grosso, 276
Fingerprint patterns, 117
Fire, invulnerability to, 133–36
Floyd, Keith, 160
Flying dreams, 272–73n
Food, life without, 153–54
Forgetting, 21
Forhan, Marcel Louis, 239
Form, disembodied, 235, 247–48, 274
Fourier, Jean B. J., 26
Fourier transforms, 27–29
Fragmentation, 75–76; Bohm's views,
49; dreams and, 63; Sri Aurobindo
and, 264–65; synchronicities and, 80
France, mass psychokinesis, 128–32